Dedicated to the chickens, who have waited so long for acknowledgement.

Stories of chickens and the Australians who love them

THIS CHICKEN LIFE

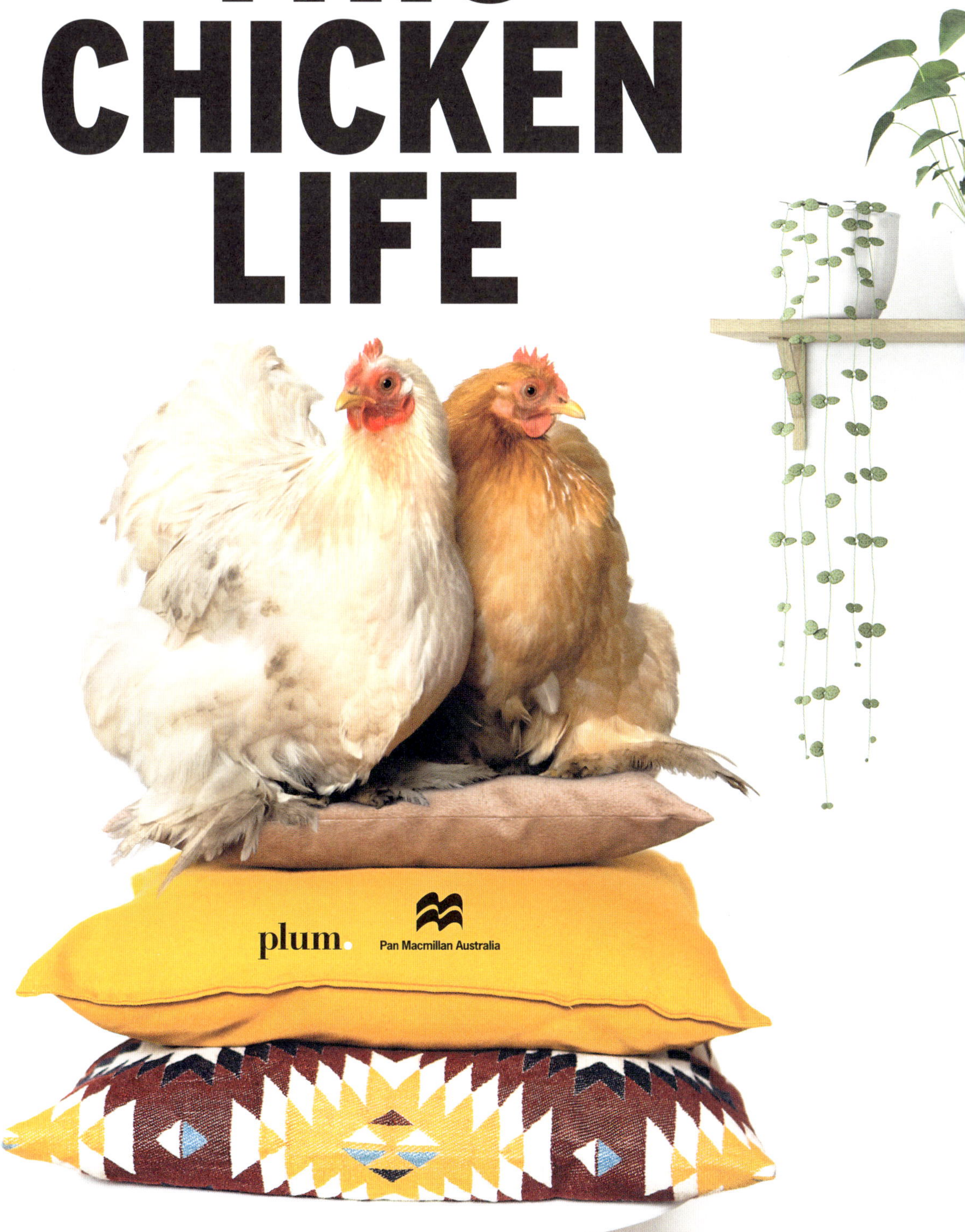

plum.
Pan Macmillan Australia

WRITTEN BY
FIONA SCOTT-NORMAN

PHOTOGRAPHY BY
ILANA ROSE

CONTENTS

INTRODUCTION

This decade finally saw things turn around for Australian chooks. The ducks, as it were, lined up. The early 2010s were a time of unironic peak hipster, and chickens ticked all the boxes: retro, practical, old-school, hands-on, showcasing lost skills, and offering the opportunity to hand-hew a coop from an upcycled car body and a packing crate sourced from your grandfather's shed. Also key: unlimited photo opportunities. Chickens *kill* on social media. Particularly heritage breeds, which were everywhere. Backyard chooks were achingly on trend in terms of food miles, organic produce, reconnecting with the land and permaculture. Capital-city real estate prices sent folk scurrying to the country, where the most immediate way to connect with a new community is to join the local poultry breeders club. We finally got serious about how dreadful caged eggs actually are. Activists were sneaking footage out from factory farms. Some of Australia's top comics starred in the 'That Ain't No Way to Treat a Lady' campaign calling for the end of caged eggs. Farmers' markets were in the ascendant, and supermarkets were no longer reflexively trusted as a food source. Schools were laying in chickens as teaching aids and friends for shy kids, and kids were going home and hassling their parents about buying free range. If you lived in the inner city a few backyard chickens became not only doable but the *vibe*. 'Why not?' people thought. 'How hard can it be? They're just chickens.'

Turns out they're never 'just' chickens.

I got my first flock of bantams – Marilyn, Missus, Stevie and Rihanna – in 2013, after months of gazing acquisitively at heritage breeds on display at my local inner-city farmers' market. Oooh, so pretty. Oooh, so unusual. Oooh, look at those fluffy bottoms. I worked in the arts, I was urban as, I had a backyard, I'd read Jonathan Safran Foer's *Eating Animals* and the idea of cruelty-free eggs was appealing: I was beyond ready. →

‘Growing veggies in your backyard and need lashings of natural fertiliser? Chickens. Social media needs a boost? Chickens. Achingly hipster? Chickens. Want to make a difference and a statement? Rescue chickens. Not connected to your community? Chickens. House doesn't feel like a home? Chickens. Need a low-maintenance companion animal? Chickens. Lonely? Chickens. Sad? Chickens. Stressed? Chickens. Could use a mental-health magic bullet? Send up the bat signal, it's chickens.’

A few friends recommended ISA Browns, praising their production-line efficiency at pumping out gigantic eggs. But I was seduced, basically, by the aesthetics and potential wow factor of their rarer relatives. There are hundreds of breeds of chicken? Some of them have fur? There's one that looks like a 1960s *Vogue* model? It was a rabbit hole I wanted to fall down.

My boyfriend, Greg, hatched a plan. He hid lumber behind the garage and built a coop for my birthday (when I saw the huge tarp-covered construction I noted drily, 'It's not jewellery, is it?'), and we drove out to regional Victoria with the empty cat carrier. I was jangling with excitement. We'd picked out three pullets online from a local chicken lady, but came home with four – in retrospect, our introduction to the fifth-dimensional world of 'chicken maths' (see page 44), a branch of cognitive dissonance that leaves you in constant denial about how many chickens you have, how many you need and how much room you have available for the always more you always need. Chickens. Beautiful and hilarious. They're just so *collectable*.

True chicken addicts are ruthless at annexing the garden, the other bit of the garden, the shed, the other shed, then the house. Greg and I were entry level. The chicken lady, Ailene, told us the girls had all been raised together, and Stevie would be lonely if she was left behind. Greg and I mentally mapped the virgin coop and thought, 'HEAPS of room'. And so it began. We now have eight girls and are on our third flock. (Spoiler alert: the fox will come. The fox will come. It will rip your chickens to shreds and you will mourn, pull up your big-girl farmer pants, and rebuild.) We've restocked twice and hatched two clutches of eggs. With eight chooks, the poo is kind of out of control, and so are the eggs. But we don't care. We get to cuddle chickens.

How did this happen? I am not a green-thumbed person. I get vertigo more than ten kilometres from a city centre. I don't have children. To be honest, when I decided I had to have chickens in my garden and life, I was responding in a Pavlovian fashion, as many did, to the Australian zeitgeist in the early 2010s. Chickens were suddenly 'it'.

They were becoming visible online as the hip accessory of choice – Tori Spelling, for example, was photographed on the red carpet with her coiffed pet Silkie, Coco, under her arm – but they were also on point in terms of sustainability. The nation collectively slapped its forehead. Chickens suddenly made sense. More than that, they seemed *obvious*. Why hadn't we been doing this all along? Where had they been? (Battery farms since the late 1950s.) Australia woke with a start and realised that the solution to many of our modern concerns was as simple as 'just add chickens'.

Reduce food miles? Chickens. Permaculture? Chickens. Going organic? Chickens. Get the kids away from their screens and back in touch with the natural cycles of life? Chickens. Food you can trust? Which isn't produced through torturing the animal? Chickens. Free eggs? Chickens (Ha ha. SO not free). Foodie? Vegetarian? Locavore? Vegan but egg-curious if the chooks are happy? Chickens. Growing veggies in your backyard and need lashings of natural fertiliser? Chickens. Social media needs a boost? Chickens. Achingly hipster? Chickens. Want to make a difference and a statement? Rescue chickens. Not connected to your community? Chickens. House doesn't feel like a home? Chickens. Need a low-maintenance companion animal? Chickens. Lonely? Chickens. Sad? Chickens. Stressed? Chickens. Could use a mental-health magic bullet? Send up the bat signal, it's chickens.

Chickens. They're the best. And they're back. Give them an inch and they'll take a mile; owning chickens is no longer a hippy affectation. In the past decade, it's become mainstream.

This is not the first time a nation has been enchanted. In the summer of 1842, bird and exotic-animal enthusiast Queen Victoria received a gift from the British Ambassador to China after the ports were opened following the Opium War. He presented her with seven thrilling 'Cochin China' chickens from the previously unpenetrated Far East. England, overcast and chill, home to dowdy, lowly and unremarkable hens, had never seen the like of these tall, upright, flamboyantly plumaged birds: everyone lost their minds. →

'I did not expect to love my chickens so much. I did not expect them to become *essential*. Buying a house without a garden is no longer an option.'

These days we'd call Victoria a 'key influencer'. She had an aviary built and spent her afternoons drinking tea, entertaining and admiring the birds. Yet more strange and beautiful chicken breeds, such as Shanghais, arrived to supplement her collection. She sent fertilised eggs to royals throughout Europe. Society rushed to emulate the Queen, and rare chooks became a fad, like Bitcoin or Beanie Babies, that swept through Britain and then America. Everyone had to have them, and they didn't care what they paid.

For about a decade, until 1855, England and America were gripped by a chicken-based financial bubble known as 'hen fever' or 'the fancy'. Equivalent to the Dutch tulip bubble or 'tulipmania' of the 1630s, when enthusiasts paid six times the average annual salary for a single tulip bulb, 'the fancy' saw collectors and speculators pay the equivalent of nearly AUD $5000 for a fancy pair of fowl. A single fertilised egg could go for $40. Bonkers, clearly, and not sustainable. By the time the bubble burst, leaving investors with egg on their faces (a high-five for this joke, please), previously exotic breeds were commonplace, and the idea of what a chicken was, or could be, had transformed.

This was a Big Bang moment for chooks, one of many since they left the Burmese jungle four thousand years ago. The reason they're the most successful bird on the planet (20 billion alive at any given moment) is, broadly, their social nature and limitless ability to serve, relax and charm human beings. They're as domesticated and bound to us as dogs and cats.

In an overheated version of Australia's current romance with chickens, the techs of Silicon Valley in the San Francisco Bay Area have their own chicken bubble going on. They've gone full 'fancy', and have exalted heritage chickens to the position of ultimate status symbol. Stressed-out Californian software engineers are spending USD $20,000 on designer coops and feeding their gals salmon and melon; they're posting them on Instagram and giving boxes of coloured eggs as on-point gifts. Since 2012 there's even been an annual Tour de Coop – a free, self-guided bicycle tour which sees chicken enthusiasts show off their deluxe set-ups and pampered hens to hundreds of fascinated visitors.

And, as happens wherever chicken fever takes hold, the mental-health benefits of chook ownership are being rediscovered and trumpeted. 'They feed your soul,' one Silicon Valley resident notes, while acknowledging that the chickens themselves probably don't care so much about the artisanal coop she's had built, complete with hand-turned trimming and antique leadlight windows. The genius of chickens is that they will flourish under exactly the amount of care you give them, but the more you give, the more you get back. Greg calls our flock, 'An aquarium that gives you eggs.' In an increasingly anxious world, chickens are ridiculously meditative.

I did not expect to love my chickens so much. I did not expect them to become *essential*. Buying a house without a garden is no longer an option. Many Australians have, of late, been taken by surprise that way, by the intensity of an unexpected personal connection. By how enriching and healing a few chooks around the place can be. We'd all grown accustomed to chickens being anonymous and uniform – no more than *things*, really – stuck en masse in battery farms and turning up deep-fried in a bucket. We'd forgotten that chickens are domesticated. That humanity had struck a deal with them, way back when, and that chooks are individual, smart and cheeky, and respond to attention and affection. That they'll even, if you raise them from chicks, hug you back.

This Chicken Life is for us, the new wave of Australians who are obsessed with our chooks, who relate to chickens every which way. There are some old-school types, for sure, who've been meeting in drafty regional sheds for decades and showing fancy breeds, but most of us had the vague thought a few years back that 'free eggs would be nice', and here we are, hooked. Naming them after our aunts or characters in video games. Attending poultry auctions for fun, coming home with two or three more, annexing another section of the garden.

Perhaps we've moved house to accommodate more, or introduced them into nursing homes as an excellent antidote to loneliness. Some of us are rescuing chooks from battery farms, others are up at four in the morning to oil the feathers of their favourite rooster before a poultry show. Still others have changed careers specifically to care for chickens, or are making little wheelchairs for chooks with a disability, or are performing comedy shows with them, or are writing while their favourite hen pecks at the keyboard. We are all spending *way* more time than we thought possible relaxing in the garden with a cup of tea and a chicken in our lap.

Just how many backyard chickens are out there right now? I'm going to guess that no one has a clue. Government figures are there for the chickens we slaughter, but if you've got half-a-dozenish hens in your backyard or are handing out fertilised eggs to your mates, literally no one knows or cares. The gestation cycle is only twenty-one days; schools hatch them by the tonne and farm them out to parents. Even breeders, perma-farmers and exhibitors only have a rough idea of their flock size, and everyone (chicken maths) coyly rounds down to the nearest ten, fifty or hundred. Because we all have more than we care to admit, because we're all addicted.

So: meet Australia's chickens, and the Australians who adore them. Jareth Bullivant, animal liberationist, who brings his rescue broiler Sephiroth to the beach to take in the sunset. Nik Round, a tree-changer advertising executive who is focused on saving a heritage breed. Summer Farrelly from Queensland, a twelve-year-old with autism, whose chickens help her connect with the world. Shane Secombe, who gives the roosters of Alice Springs a second chance in life at the prison. Jennifer Birkin, an Adelaide comedian who performed a show with her chickens called Crazy Chicken Nerd. Instagram star Mandy from Queensland, who crochets chicken hats that sell like hot cakes on Etsy.

Ilana Rose set out with me to capture 'the fancy' in action; her photographs are stunners. She was chicken agnostic and hadn't expected to fall for them, but you should never underestimate the allure of a chook.

For the first time since battery farms were created in the name of yield and efficiency, Australia's relationship with the chicken is undergoing a major transformation. People have been reunited with chooks. And chooks – finally – have been lifted back up where they belong. ●

DISCLAIMER

I AM NOT A CHOOK EXPERT

I mean, sure, I am compared with people who don't know anything about chickens. To my friends who know me as a cabaret director, writer, DJ, performer, average swimmer and ranga, I have metamorphosed into an extreme poultry enthusiast/bore. I am always ready with a hilarious chicken fact or adorable photograph of one of my eight adorable chickens being adorable. Ask me anything and I have an answer. Fastest gun in the run! *Pew pew!* I have opinions backed by lived experience and poultry groups and books. And the more I learn the less I shut up about it, because chickens, it transpires, are one of the central planks we have for saving the world, so if you were to label me a chicken proselytiser you would be correct.

But. My knowledge is limited: I am no vet or breeder or permaculturalist or third-generation farmer. I'm not even a proper hipster. I'm an inner-city creative with few prospects who, for once, caught the zeitgeist and got chickens on the upswell. It's important you understand that although *This Chicken Life* contains my opinions, tips, experiences and hot takes, it's not a 'how to', or a complete guide to anything. There are *so many* of those books, all full of authoritative information; I own several and I recommend you do too. It's important to read widely and amalgamate as you go, as there's always something more to learn. Chickens are a marathon not a sprint. This has become particularly apparent to me as I've met so many brilliant, committed people while writing this book. Also, there are different opinions on everything. *Everything*. Please don't choose my hill to die on.

Just this week, for example, I discovered that you shouldn't feed hybrid chickens such as ISA Browns solely on scraps and grain. They're genetically engineered to require more protein and can't survive on what used to be regular chicken feed. They need manufactured pellets to a) pump out the eggs and b) not weaken and die due to pumping out said eggs. Huh. Essentially, thanks to human intervention, the vision of a happy brown chook popping out an egg a day on scraps from the kitchen is nostalgia and we are currently living in the Matrix. But if you didn't *know* that, how would you know?

I've also read that you should 100 per cent give your chooks scraps in the morning, so they'll scoff them before filling up on grain, which ensures a mixed diet and keeps vermin at bay. Also, that you should 100 per cent *not* give chooks scraps until the afternoon to make sure they fill up on pellets or grain first and thus ensure they're getting all their nutrients. (I've decided to alternate.) I've read that pellets should make up the entirety of a chicken's balanced diet *and* that pellets are the devil and it's only through foraging and scraps that the chooks will be happy and the eggs tasty.

The answer, I reckon, is a little of column A, and a little of column B. Chooks pick what they fancy from what they're offered, and so should we. The only thing I'm certain of is that I know just enough to know that I am a long way from offering definitive advice. (FYI, I feed my chooks Red Hen scratch mix and all manner of scraps. They are very keen on cheese. I believe this diet is known as 'spoiled rotten'.)

How to best care for chickens is contested science depending on your approach, your circumstances, your needs and your relationship to chooks. It's not for me to adjudicate. Some people, for example, will take their ailing feather baby to an avian vet and drop a couple of grand, others will simply dispatch the bird. There is a rich spectrum of options between these two extremes. None of them are wrong, none of them are right. It just depends.

I sit somewhere in the middle, working on the theory that if a chook presents as sick – that is, wilting like an unwatered pot plant and perhaps sneezing – separate it immediately because contagion is real. Then do a happy dance because you have a house chicken for a week. Stuff it with mealworms, corn and other tasty treats, and order antibiotics from the vet. I figure it will recover or die, and so far in my experience they've come good. Is this advice?[1] After three flocks and a number of health crises, it's where *I've* ended up. Many roads lead to Buddha, and the only thing I'm clear on is that they're in your care, and curious, smart, trainable and genuinely affectionate, so treat them accordingly.

This Chicken Life is a book of true stories. It is intended as an emotional roadmap, not a practical one. There are tips and opinions, but mostly you will meet some of the Australians who love chickens, and learn why. These are epic tales, bristling with heart and death, humour and vulnerability. Maybe you'll find that it's okay – necessary even – to love chickens. And if you already do, you'll find yourself reflected back. ●

1 NO. Not advice. Pay attention.

MY FLOCK

My current flock consists of eight heritage hens bustling around my inner-city backyard. They're all bantams, which is chicken jargon for 'small'. Think boxing, and bantamweight fighters. But laying adorable, tasty, teeny eggs. Bantams are, IMHO, the best chickens to have because you can sneak more into the same space, and once you start owning chickens, that becomes a priority. Technically, according to the local council, I'm allowed up to six chooks, but I've done the chicken arithmetic in my head, and six full-size chickens are equivalent to about eighteen bantams, so, pfft, we're good. Being chicken maths, of course, this is not legally binding, so if anyone from my council is reading, I have six chookens. But two of them are broody at any given time, and aren't laying, so, again, don't really count.

I'll introduce you in pecking order.

Priscilla and Phyllis are the white Silkies, all fluff and foof and getting-under-footness due to zero peripheral vision. We named Priscilla, the largest and fluffiest, after the film, for her sheer high campness, and Phyllis for the American comedian Phyllis Diller, because when she's caught in the rain her pompom spikes into a fright wig. They have iridescent green earlobes, black skin and black bones, and if you manage to photograph them from underneath, peering up at their angled, alien faces framed with white fur, they bear a remarkable resemblance to David Bowie in *Labyrinth*.

Next is my Australorp bantam, Lally, a stoic chicken and good layer, with black and oil-slick-green feathers, named after the Melbourne playwright Lally Katz for no reason other than that I liked the meter of 'Lally the 'lorp'. Lally used to be trusting and the first to bustle up for treats, but I dipped her legs in petrol to deal with a nasty case of scaly leg mite, and she's never forgiven me. Now she eyeballs me suspiciously, at arm's length, snatching perhaps one corn kernel from my hand and retreating. Desperate for forgiveness I try to lure her in with melon pips (her favourite), but no dice. →

Fiona pictured with Dora and Val, Rosecombs.

I fret that she's lonely. She's the only one of her kind in the flock, so has no one to hang with. This is my third flock, thanks to the predations of Ms Fox, and I've discovered that chickens are *totally* racist. They band together according to a strict hierarchy of breed, then size, then colour, and exclude accordingly.

The remaining five are Rosecomb bantams – tiny, flighty and bossy. Chicken Napoleons, the lot of them. Their position relative to each other in terms of pecking order is fluid; because they're all big personalities, there is a lot of jostling. The two black Rosecombs, sisters Violet and Joan, are the eldest of the flock, coming into their third season. Violet, highly strung and in the habit of suddenly shrieking and belting across the garden as though pursued by bats, was named after the colour of her earlobes.

Joan, one of my favourites, had a single white dot on her tail when we first got her, so she took the name of Victoria's first female premier, Joan Kirner, who was portrayed in political cartoons wearing a polka-dot dress. Joan has a chronic and hilarious case of 'overbeak', so her top beak needs regular clipping or she looks like Gonzo from the Muppets. It's tempting to leave it, as it adds personality, but a chicken's beak is its everything: mouth, hands and weapon. She's prone to epic moults – bald head, chunks of feathers missing, others hanging on like wobbly baby teeth at wild angles – which give her a crazed, ratty, starving look, like a taxidermied vulture nibbled by mice. Frankly, she can be a sight.

Her sister, Penny (named after the head of my writing department at uni, Penny Johnson, because she was lovely and the flock's first top chook), literally fell off the perch recently. This is awkward and why you shouldn't name chickens after friends. Now Joan's lost an ally and is having to assert herself or lose her place in the pecking order. I'm putting down protein – cans of Aldi pet food – to help her grow her feathers back, but instead of chowing down, Joan's hurtling around, warning the young ones off the sweet cat-food action. 'Save your energy!' I want to yell. The tatty little duffer.

The chickens make me feel godlike. I have the power of life and death over my flock, and yet have to watch helplessly as they make bad choices.

Finally, there are the youngsters. The pullets. In chicken terms, pullets are what Jane Austen would deem 'young ladies not quite of marriageable age'. They're three Rosecomb teenage girls: one 'splash' (white with flashes of black) and two 'blues' (they're grey – grey chickens are called 'blue'; there is no consensus on why). They form their own sub-flock, with their own hierarchy, like a pack of teenage girls. The fattest of the blues is Dora, named for the explorer. She is characterised by buxom good health and a tendency to be first out the run door to free range, shadowed by Blondie, the splash, who has a wide ring of scar tissue around her neck and is thus the pirate queen of the second tier.

Blondie was named androgynously as a chick too young to sex, while she spent a week inside the house recovering from being savaged by Priscilla. Who has issues. Mostly that when she sees a baby chick, a red mist descends and she tries to peck its head off. Which is unfortunate, because she longs for babies, as all Silkies do, and goes broody in a heartbeat. A lot of breeds have had the maternal instinct bred out of them, and are all like, 'Screw having kids, I'm going to the disco. Laters.' But Silkies are still slaves to their hormones and love nothing more than spending three weeks in an egg-sitting coma.

As sisters, Priscilla's and Phyllis's biorhythms synchronised, and they went broody together, sat on eggs together, snuggled in an expectant, fluffy clump together. Adorable, except none of the fully developed chicks made it out of their shells, because Priscilla heard them peeping and PECKED THEM TO DEATH. Priscilla has had her hatching privileges rescinded.

And then there's Val, my favourite, who doesn't fit in and is so unpopular my heart lurches. That was me at high school, and I'm afraid I project. Val's a teeny bit Not Quite Right. She's smart and all, but as a littlun couldn't find her way into the coop to save herself. She'd watch the others troop through the door at sunset, then she'd fly at it, miss and, finding herself on the roof of the coop *again*, would heave a huge sigh. She likes being high, and has found a perch outside, my theory being she doesn't want to be inside at night with all the Bitches. We scoop her up after dark, share a cuddle and a chat – she's very chatty – and put her to bed. Don't want her getting cold.

Every flock. Every flock is a soap opera. ●

ILANA ROSE
PHOTOGRAPHER

While not currently a chicken person herself, *This Chicken Life* photographer Ilana Rose knows a thriving subculture when she sees one, and has immersed herself enthusiastically in Australia's burgeoning love affair with chooks. Passion is passion. That said, Ilana is well positioned to appreciate the hypnotic hold which chickens can exert on their people. In her final year of school biology there was a duck-hatching and imprinting project – the kind of dubious experiment which is frowned upon these days – and Ilana was bonded to Dibs, a little drake named after a clinical psychology book called *Dibs in Search of Self*.

Dibs was taken into Ilana's family fold down on the Mornington Peninsula, given a home in the shower recess, and had quite the impact. Ilana developed a propensity for bringing home stray animal anythings, and her artist mother picked up the poultry bug and ran with it. Shoshanna, who had a garden studio, obtained about thirty more ducks. And then, naturally, came the chickens. Oh boy, did they come. At the flock's height there were up to eighty chickens and ducks eating fallen fruit in the garden. Shoshanna bonded with the chook at the bottom of the pecking order. Blimp (the gimp) was a Rhode Island Red with a deformed beak and a bung foot. Rejected by the rest of the flock, he became her muse; he had a special little cushion in the studio, and appeared in many of Shoshanna's paintings. Ilana's still got one of them. ●

THE ONE-WINGED ANGEL

JARETH BULLIVANT

Animal liberationist, bird lover, fantasy gamer

Lives: Mordialloc, Victoria

Pictured with Twistie, a rescue broiler

'Sephiroth weighs 5.2 kilograms. The vet put him on the scale and said, "That's a dog."'

NUMBER OF CHICKENS:
Five (four hens, one rooster)

BREED:
Cobb hybrid (rescue broilers)

NAMES:
Aries, Cloud, Cooper, Twistie, and the rooster, Sephiroth. 'He's named after a strong, proud, masculine character in the video game *Final Fantasy*. Sephiroth's a villain who was lied to and used and, in a way, that's how the chickens have been treated too.'

OTHER PETS:
A rescue pigeon called High Wind

Jareth Bullivant's selfless devotion to birds is evident from his sparsely furnished, 1950s one-bedroom flat. At a glance he could almost be squatting, sleeping on a double mattress on the wooden floor of the living area, a pile of chicken poo swept tidily into the corner. There's no TV, but a small screen sits next to the mattress for gaming and video-chatting with his English girlfriend Mel, who runs a bird sanctuary in the UK. There's a sheet-sized summer-sky-blue fabric mandala tacked to the wall behind the bed. It brings out the colour in his eyes. A pigeon swoops possessively around the room.

'High Wind's not used to visitors. She wants to know what's going on,' he says.

The only bedroom, complete with ensuite and actual double bed, is given over to Jareth's four chunky rescue hens – Twistie, Cloud, Aries and Cooper – who sleep resting their dense breasts up against the foot of the bed frame. It's rare to see adult broiler chickens at all, let alone hanging out on a handsome dude's bed. They're bred to pack on muscle and be slaughtered at six weeks. At five months old they're massive and still.

It's rare to see adult broiler chickens at all, let alone hanging out on a handsome dude's bed.

'They kept growing until they needed their own room,' says Jareth, thirty, who rescued them as part of a protest at an abattoir. 'I don't mind. I like sleeping on the floor. I prefer it, to be honest.'

On the Facebook group The Politically Incorrect Vegan, in the first photograph in their post about the Direct Action Keysborough Star Poultry protest, Jareth is dead centre. It's hours before he went home with a car full of spontaneously liberated broilers, and he's clutching an arterial blood-coloured 'Close all Slaughterhouses' banner. Jareth was one of about fifty activists who staged a peaceful protest outside the abattoir in December 2017, in response to an ABC *7.30* exposé, which broadcast footage of chickens having their feet sheared off by machinery before being boiled alive. Staunch, slightly fey, with black boots and long blond hair well past his shoulders, Jareth is elf lord Legolas made manifest. As a gamer, this is probably not a coincidence. He certainly has the air of being engaged in a supernatural fight against an implacable enemy. →

'I don't like to use the words "fight" or "battle", but that's what it feels like. An ongoing battle. What I want is for people to understand that the separation between humans and animals is an illusion. What those animals go through is so horrible. There's a part of me that always wants to say, "I don't like eating animals because I think that it's wrong," but if you say that to people, nine times out of ten they think you're an arsehole for making them think about things and feel bad.'

Jareth, who has a large dark tattoo of two crows on his neck, adores all animals, all birds, but he's closest to his magnificent rooster, Sephiroth. There is mutual adoration. Seph weighs in at 5.2 kilograms, and walks gingerly due to a case of bumblefoot, an infection which often affects the feet of heavy birds. His neck wattles are the size of saucers, the colour and texture of ripe testicular strawberries. As Jareth talks he massages Seph's dangling lobes, and the rooster relaxes, ecstatic, and drifts in and out of sleep, snuggled safe and tight in Jareth's heavily inked

'I don't like to use the words "fight" or "battle", but that's what it feels like.'

arms. Jareth can't turn the world vegan, or save every distressed bird from the abattoir, but he's having a red hot crack at making one rooster very happy.

As the moon orbits Gaia, Jareth's energies revolve exclusively around Seph, High Wind and the gals. He works, somewhat grudgingly, in a local cafe part-time to make money for vet bills, chook food and rent. 'People at work ask me, "What did you do on the weekend?" They're always out, doing this and that. And 95 per cent of the time my answer is, "I spent time with my birds. Played some games. That's it."

'I just like to care for animals. The only reason I go to work is so I can be here with them. I'm providing, the same as any mother and father.'

The long game is for Jareth to join Mel, and her sister, in the UK and launch their own sanctuary. 'They're carers. Very inspiring. I've learned all I know from them. It started when they found a baby sparrow, Mortimer, and she lived with them for nine years. The most incredible bird. →

'Now people ring them, saying they've got this bird or that. They always say yes. They've got so many birds. They don't do anything illegal – they go to chicken rescues and then they rehome them.'

While he's planning and saving money to travel, Jareth is in Mordialloc in his flat by the seaside. Seph, the more proximate of his affections, sleeps in a pet carrier near the mattress. Two reasons. First, if Seph's in with the ladies he mounts them, and his fave gal, Twistie, has a nasty gash in her side from his romantic manoeuvring. 'He's too heavy for them,' Jareth explains. Second, when Seph wakes early and crows, Jareth, mindful of the neighbours, can leap off the mattress and intervene. 'When he crows I go to him, put my hand under his neck, and he nuzzles in and goes back to sleep.'

The flock gets to be outside when Jareth's home, either on the communal grass or hanging out in the yard. He's fenced off his own section with salvaged pallets and built a coop for the flock (which they outgrew), and he takes them one by one to the beach to soak up the spray. When they were smaller and lighter, he bought a cat carrier backpack with a see-through dome window, and knapsacked them across the road.

'I take them down at sunset, sit on a rock, and they sit on my lap. We watch the sea and the sun going down. I wanted them to experience sunset at the beach, basically. Why not? It's lovely. Most of us think it's lovely, why shouldn't they have that experience as well?' ●

LADY MACGYVER

MIRANDA BOULTON AND CORI HAWTIN

Dynamic duo, animal rescuers, makers of chicken wheelchairs

Live: Murrumbeena, Victoria

Miranda pictured with Pepper, a Silkie

'We call ourselves "foster failures" because we don't give anything back.'

NUMBER OF CHICKENS:
Seven (hens)

BREEDS:
Araucana, ISA Brown, Leghorn bantam, Silkie, Silver Laced Sussex

NAMES:
Scratchie, Sunday, Custard, Strudel, Nellie, Slippers and Pepper

OTHER PETS:
Two dogs, a German wirehaired pointer cross and a labrador, Pip and Lawson; two horses, Henry and Willow; four alpacas, Floyd, Oscar, Nugga and Brandy; and twenty-four sheep, Dolly, Abbey, Banjo, Frankie, Ben, Noah, Lizzy, Wilbur, Clove, Crumpet, Jamie, Olive, Lily, Clancy, Poppy, Missy, Penny, Sprout, Charlie, Brownie, Cocoa, Winnie, Gypsy and Stelly

YAMBEE

There's a touch of MacGyver about Miranda Boulton. She's resourceful. Handy with tools. A problem solver. The neighbours lost their flock to foxes, which means you're next, so she fitted her three chicken coops with auto-closing doors rigged from old car aerials and a couple of laptop batteries, with timers plugged into the mains. 'You just hope you don't have a power failure.' She's installed cameras (plural) inside each coop, so she and her equally rescue-obsessed partner Cori Hawtin can check and see that everyone's tucked in. 'It's just like baby cam, isn't it? For our babies,' she says drolly. 'It's nice to see who's sleeping with who, what's different. Or see if someone's not gone up to roost – maybe something's wrong, and we need to look at her in the morning.' She glances up from her phone. 'To be fair, we are extreme.'

Evidence of Miranda's ingenuity is everywhere. In the child-proof fencing that keeps their boisterous pup from the chickens, in the stacks of salvaged materials she's accrued to build a tiny house (including a chook claw-foot bath; she's a total hound for hard rubbish), and in the portable coop she's built into an old trailer (named 'The Pull-et', which, by the way, is an excellent chicken joke) that they use to take the chooks to their holiday house on the Peninsula. 'Yes, they holiday with us,' says Miranda. 'They know where they're going, and they're like, "We love it down there!" It's free range, a different garden, one they can rip up.' They both laugh.

'It's their holiday house too,' Cori says, grinning. 'So if you thought we were crazy before . . .'

Let's just say *ardent*. The public-facing component of their lives is eminently respectable – Miranda's the Out of School Hours Care program coordinator at Murrumbeena Primary, Cori's a chiropractor – but beneath the conservative veneer are a wild pair of do-gooders who prioritise animals above all else. If they're not at their animal sanctuary, Willow Haven, tending their twenty-four rescue sheep or giving their rescue pony a chiropractic treatment, they're plotting secret business through their rescue group, Dreamer's Hen, or feeding screaming lambs milk at 3 am, or scanning ads for sheep being offered as 'suitable for the BBQ' so they can hitch up the trailer and go save them from the bad guys. It's a high-adrenaline, low-sleep caper when the game is on.

Their shared love for animals is the foundation of their relationship, and something they've grown into. They started innocently enough, with three chooks, as Cori had always been chicken curious, which of course led to more chickens. Then they became involved in a rescue group, and another rescue group, and then Miranda's parents bought a farm in Gippsland, which meant all of a sudden they had ten times more space to bring home sheep and lambs, and look, they're the first to admit they're a bit out of control. 'We've spent probably as much money on our animals as parents might on their children,' says Miranda. 'My whole wage, literally, goes to caring for them, feeding them and spoiling them. You do it because you love it.'

Horses are particularly expensive to care for. But one of their rescue hens, a chicken called Stormy, racked up $4000 at the vet. 'That wasn't in one hit,' Miranda notes. 'That was over months of treatment and medication and surgeries. But when you look at the end, yup, that was a four-thousand-dollar chicken.'

'But we'd do it all over again,' says Cori. 'We would never go, "Okay, we love that animal to a dollar figure, and when it hits two thousand dollars it's over." We do the best we can.'

'A lot of people go, "You're mad,"' says Miranda. 'My friends look at me and say, "Oh, you've got a chicken and you're taking it to the vet? Really? The vet? A chicken?" I don't even mention it's an avian vet. They'd be like, "A specialist!" Yes, a specialist. But you know, they're our kids so they get the best of care.'

They go above and beyond, plain and simple. Miranda is also the founder of Chicken Therapy Chairs, a home business which makes and ships wheelchairs for post-surgery or incapacitated chooks. You could call it a side hustle, except that it's not priced to make money; the chairs simply cover their costs. Miranda can often be found outside, spray-painting PVC piping bright primary colours, or inside, packaging orders late into the night or sewing the slung cloth hammocks (with leg and poop holes) for the wheelchairs. 'They need to get sent quickly, because people want them *now* because their chicken's sick *now*. →

'They're desperate, and it's five to ten days' delivery for some countries.' There are four standard sizes (including a teeny weeny 'chick'), but nothing's standard when it comes to a chook with special needs, and chairs often end up customised. 'There'll be a chicken with one leg, or one with wry neck that needs a lower rail at the front to reach the food bowls. Another chicken won't have wings. Every chicken has a story.'

The demand for chicken rehabilitation chairs is high, and rising. 'Work hours aside, and the time we're at the farm, every spare minute is doing this.' Miranda sells up to ten chairs a week, over five hundred so far, with word spreading fast via online networks of animal sanctuaries and chook lovers, and break-out news stories. From a media perspective, it's a quirky story on a slow news day (chicken wheelchairs? LOL), but the therapy chairs address what's become a genuine issue now that chooks are being viewed as pets rather than just livestock. It's only obvious in hindsight that ill companion chooks will have after-care requirements. How *do* you physically rehabilitate a chicken with a broken leg? Or one that's had a stroke? And how do you give quality end-of-life to a rescued broiler chicken with leg and back deformities? On the most basic level, chickens need their bums off the ground, otherwise they sit in their own waste and can't forage. Everyone's been playing catch-up, vets included.

Miranda's call to arms was one of the hens they rescued, Neptune. Poor little chook. She'd been rough-handled in a battery farm depopulation process, and had two broken wings and two broken legs. 'Obviously she was in a very bad way. The vet pinned both her legs in surgery, and realigned her wings and bandaged them in, and there was just no way of supporting her.

The vet had her lying on her side post-surgery, but it wasn't comfortable because of her wings. They said to me, "You build things. What can you come up with, because at this stage we're literally thinking of hanging her in a stocking in a doorway, in a Jolly Jumper."' Cori and Miranda arrived to see the hen suspended by pegs in a kid's wiggle car. 'It was ridiculous.'

Miranda comes from a background in plumbing and irrigation, so her mind went straight to PVC pipes and fittings and elbows. She threw something together, attached food and water bowls to the front, and added wheels for novelty. 'I thought, "Ha ha! I'll see if this flies!" And it did. It worked!' Cori, as a chiropractor, advised on adjustable height straps for the hammock seat. The vet was blown away.

For a while Cori and Miranda used that one chair for their chooks, then a connected American chicken lover saw a photo on their Facebook page, insisted Miranda set up a website, linked to it and social media – as it sometimes does – happened. 'It just went *boom.*' Their biggest break was Californian Darcy Smith, who runs a high-profile animal sanctuary called Funky Chicken; she used one of the chairs to rehabilitate her rooster, Roo, who'd sustained a head wound in a fight. 'He basically went from paralysed to walking on Christmas Day, conveniently enough,' says Cori. There was a storm of media attention from *The Project* at home to ABC and NBC in America. 'Devo we didn't get on *Ellen*,' Miranda says.

The chairs aren't for every chook. Some are too flighty. Some just don't take to them. They're certainly not meant to be lived in. But, coupled with actual rehab, the chairs are helping chickens relearn how to walk. →

'We've spent probably as much money on our animals as parents might on their children.'

With the hammock supporting their weight, and the wheels providing mobility, they can re-establish the different phases of gait mechanics. Many orders come from America, where frostbite causes chickens to lose their feet, and where bushfire events in California have burned them to stumps. The stories of need are compelling: one chicken with a broken leg after being chased down by a dog; another lame from scaly leg mite; the broilers (so many) with collapsed legs. 'The broilers are the most heartbreaking,' Cori says. The chairs work, though. Animal sanctuaries and vets are turning into big customers. 'They order one in each size,' says Miranda.

Miranda and Cori see Chicken Therapy Chairs as part of their service to the community. 'You'll get an email back saying, "My chicken can finally sit upright and eat!" They're over the moon,' says Cori. The chairs bring definite quality-of-life benefits, some of which are social. A chook in a chair is one that can be outside with its flock. In some ways best of all, the therapy chairs have opened Miranda and Cori up to a community of people who are like-minded, both here and internationally. After years of good-naturedly swatting away judgement, it's been wonderful to realise that there are many others who share their passion. 'It's huge. I never realised, but there *are* people in Australia who love their chickens like we do. Other people recognise these little individuals, and that they're part of the family. Suddenly we're going, "Oh! We're not crazy!"'

Still, a little bit of crazy is the fuel in their tank. They have to play it prudently at work, so at home they thoroughly enjoy the roles of extravagant animal champions. 'We walk into our country vet, because we're probably the only people who wouldn't shoot a sheep that's sick. They go, "You again. What have you got today?"' says Cori. 'We open the boot of the car and there's a sheep and lamb, and they're just like, "You're mad." And then we had a sheep and a lamb down here, and took them to our vet in Bentleigh, and the looks on people's faces in the waiting room when you walk in with them. "Really? In *Bentleigh*?"' Miranda hoots. 'So we're celebrities down here.' ●

BINGARA

THE ADORATION

COSTA GEORGIADIS

Host of the ABC's *Gardening Australia*, sustainable living advocate, landscape architect, positivity machine, chicken messiah

Lives: North Bondi, New South Wales

Pictured with Flighty and Awesome, Australorps

'I would respect chickens more than a percentage of the population. They're not exploiting people. They're not miners, just sucking things out. Chickens contribute. They put back. They do the job of regeneration better than the way most humans are approaching it.'

NUMBER OF CHICKENS:
Two (hens)

BREED:
Australorp

NAMES:
Awesome and Flighty

OTHER PETS:
150,000-odd worms, 10,000-odd native bees and a few thousand honey bees

Most of us would blush to be looked at the way Awesome is drinking in Costa Georgiadis. She's eating him with a spoon. Costa's holding her warm against his chest and laughing, backlit in his unkempt garden by a nimbus of sun, and Awesome, an Australorp hen possessed of stupendous good health and a radiant red comb, is either love-struck or in a religious trance. It's possible she's just vibing on the exceedingly high regard in which she's held. Chickens are some of Costa's favourite people, and when he gives a chook attention, he gives it 100 per cent. 'You hold them on your heart,' he says, 'and when you're calm, they're calm. Because they connect to your heartbeat.' Costa, grounded as a mountain ash, radiates serenity. If Awesome's connected to Costa's heart, it's no wonder she's awestruck. He adores chickens. They're his secret weapon.

Costa Georgiadis – he of the magnificent 'two owls and a hen' Edward Lear beard, the charismatic host of *Gardening Australia* and saxophonist with shambolic funk outfit the More Please Orchestra – worked out some time ago that lecturing people on their bad habits is not the road to change. No matter how urgent the environmental situation, proselytising is a mug's game. 'Whacking the mirror in front of people just doesn't work. In the early days, my exuberance and energy were over the top. Now I pare it back. I just place some road markers out there, stir the current.'

Far better to go into your dance, entertain and delight, and pique someone's curiosity with, say, your two chooks free ranging on your edible verge garden; then, *bam*, an intrigued bystander is on the sustainability path. Or take your favourite hen to a lecture and perch her on the lectern and, maybe, while you're roaming in the crowd, she clucks into the microphone and charms the crowd's pants off.

It's a shrewd strategy. Costa, a gifted physical comic who once worked as a clown at Darling Harbour, radiates welcome, warmth and sincere wonder about the world, and couples it with a connection to soil and soul. He also rocks an unironic steampunk gnome costume. He's a goddamn delight, and adults and kids alike are drawn to him like moths to a torchlight trained on a bedsheet hung from a Hills Hoist. (Which, FYI, is a great way to gather insects as a treat for your chooks.) Never underestimate clowns; they have serious intent.

For Costa, of Greek heritage, exposed as a child to his grandfather's lush and productive backyard market garden in Randwick, chickens are a necessity and a no-brainer. 'They were in my life as a kid, in the background, part of my DNA. Now I have my own chooks, they're front and centre, and I'm dropping that DNA: laying it down, sprinkling it around for others to pick up. Chickens were part of the family, always part of the village, yet, like everything else, they got commodified and separated. They've been distanced from us, and we've lost respect for them. Without the respect it's just a piece of "something per kilo wrapped in plastic". We need them back.'

You're probably familiar with Costa's verge. The edible share garden by his Bondi back door is the narrative epicentre of the sustainable community he's building, and the focus of weekly stories on *Gardening Australia*. He's always elbow deep: making a worm farm out of an old bathtub, helming a group planting, enthusing about what's been grown, demonstrating the 'chicken tractor' he's made from an old dining-room table, illustrating how chooks turn the soil over, how easily they can be incorporated into suburban life. 'You know, if you feed 'em grain it makes their poo nice and dry and you can scoop it up off the tablecloth. No drama.' Costa's master plan is immodest: to wrest our food destiny back from big agriculture.

'The future of food is local, urban. It's an incredibly powerful process to bring back control to *us*, to bring food back into suburbia. On a balcony, a nature strip, a small backyard. We tend to think that, you know, Big Ag is going to save the world. It's such a limited lens. It's not even the case. At present, 70 per cent of the world is fed by women growing small, small allotments. And the difference of growing something yourself! Tasting the difference! Having control! The future of food is small. It has to be. It's like when the Chinese said "no more" to our waste, and suddenly we have to face that there is no "away".' →

Chooks are key players here. Costa rhapsodises about the physical permaculture work they do – making compost out of scraps, ensuring there's no waste, turning over and fertilising the soil, eating weeds, providing eggs – but it's more than that. Costa sees Awesome and Flighty as partners and, like him, they're non-threatening and magnetic. Just by being their adorable, fascinating, fluffy-butted selves, they draw attention.

'They're an amazing portal. I only have to have the side gate open and local kids – you know, fifteen-, sixteen-year-olds – are screaming out because they've seen the cameras. They say hello, then they ask how the chickens are. A family from over the street just walk in and say, "Oh, Costa, the kids want to come and visit the chickens." Random people get excited walking past because they see a chicken. They're in the city, and there's a chicken!

Putting a chicken out there in an unexpected, public setting, it puts these little incendiary cues in people's memory. It's about disturbance. The more you can disturb the status quo, the more you can put something out of context, that's when you get cut-through, without it being a donk on the head. The chickens become a key to dialogue that I don't have to activate.'

Winner of the Most Popular Presenter Logie in 2019, Costa is, nevertheless, the inverse of a TV personality. The door to his garden is often open; his shareable food verge is there to be picked; he leans in to people rather than locking himself away. One of his proudest achievements is the neighbourhood food-scraps drop-off. Locals deliver their food waste – a couple of wheelie bins' worth every week – to Costa's door, which explains the robust health of Awesome and Flighty (and Costa's metastasising worm population). →

Costa's philosophy is to do a lot with a little. People in the inner city don't have access to much land. Doesn't matter. Costa's used to working within constraints himself: his backyard is 90 per cent swimming pool, a legacy of different times. It's a good source of algae ('Excellent for mulch!'), and he's turning it into a water garden via segments on *Gardening Australia*. The house was his parents', and he lived there with his father until Stan died in 2017. It's a big, chaotic, rambling place, and Costa would be rattling around in it if he weren't away at festivals and events every weekend, saving the world. When he comes home, the chickens provide company.

'Even from the moment I open the door, you can hear them. They bring a sense of companionship to the household. Much like a dog. I turn up, I go out to see them, I'm talking to them even if it's late, and they'll hear me and come out as if to say, "What's going on?" I'll bring them a little evening snack. I s'pose it's a belonging. It's a presence, you know. It means someone's home. And that's, that's something that you just can't get, you know, looking down at a phone or looking at a screen. Chickens have changed my life in ways I couldn't have imagined.'

Out on Costa's verge there's (another) rescued bath waiting for its next life as a worm farm. He jumps in for a photo, no hesitation, holding Awesome and Flighty. There's a single shaft of something green, tall and weedy-wheaty between Costa and the lens, and it's uprooted for the perfect shot. Costa reacts with urgency for the first time, a hand shooting out with alarm. 'Save that weed!' he says. 'The chickens will like that.' ●

CHICKEN MATHS

A HAPPY AFFLICTION

A post in the Australian and New Zealand Facebook group Backyard Poultry:

'I asked my Google Home Mini "How many chickens should I have?" She replies with "Bla bla bla recommends 3 chickens per member of the household so if there are 2 people you should have 6 chickens."[1]

'My wife says, "Whoops, we have way too many chickens."

'I reply with "No. It works like this: 1 person = 3 chickens. 2 people = 6 chickens. Now for every chicken you have you can have 3 more chickens. I mean like . . . they are members of the family and all."'

Every comment that followed this post was a variation of 'Yes! Now I can get more chickens!'

•••

In 2015 an American-based website called The Happy Chicken Coop held a survey, asking experienced chicken owners to share the one thing they wished they'd known before getting backyard chooks.

The overwhelming number-one response wasn't 'how to deal with predators and rodents' (number three), or 'common chicken illnesses and how to treat them' (number two). It was 'chicken maths'. 'The one thing I wish I would have known before getting them was just how addictive they are,' said Jennifer S, her full name redacted to protect her family. 'I started out with five chicks in July last year, and today I have almost thirty chickens.'

Consider yourselves warned. →

1 Two people do not 'need' six chickens. This is more chicken maths bullshit.

Chicken maths is a little-known offshoot of quantum physics which runs on hope and cognitive dissonance.

It is the dark secret shared by every chicken enthusiast, the affliction which unites us all: the unquenchable desire for more chickens. A happy affliction, I hasten to add, but an unavoidable consequence of obtaining chooks. They are the crack cocaine of pets. 'Oh, I'll just get two. For the eggs,' we think, and the universe laughs.

The signs of chicken maths at work are twofold:

1) You are unwilling to admit to, or count, how many chickens you have.

2) The size of your flock, like the universe, is constantly and inexplicably expanding. →

CHICKEN MATHS: THIS IS HOW IT WORKS

1

You decide, as per the aforementioned scenario, that you're going to get a couple of chickens for the eggs. You go to the market, or a farm, or a store. The chicken seller explains that you need *three* chickens, because one may die, and chickens are social and you can't just have one chicken; it will get sad and die. This is all legit and not a weird Ponzi scheme. You come home with three chickens.

Congratulations! You're already on the slippery slope of chicken maths and you haven't even got them out of the car.

2

One of your three chickens dies. Or turns out to be a rooster and has to go back to the store/'live in the country'. Either way you're down to two. Now that you're used to three, two doesn't look right. You've watched them interact, you know that three is the minimum optimum dynamic, and you've already worked out that they have unique and hilarious personalities. You miss Molly/Nando/Boadicea. Also, while the regular brown hybrids are great for laying eggs, you're now in the world of chickens and know of some spectacular-looking heritage breeds. You're officially chicken curious. You go back to the farm, or chance your arm at a poultry auction.

Ooh. *So* many gorgeous chickens. You know not to buy one lone chook; they need a buddy, particularly since you're introducing them to an established friendship pair. You lose your mind a little and come home with two or three new gals. Say, some Buff Orpingtons, with their adorable fluffy butts. Or some Australorps, with their oil-slick green-sheened feathers. Or some Silkies because they are made of cotton wool and kittens. You now have five chickens.

3

The coop you provided for two chickens is too small. You build a bigger coop. There is room for more chickens now. Also, the Silkies keep going broody, so they're not producing many eggs, but wouldn't it be *lovely* to hatch your own chicks? Also, there's a garden corner over there that's not doing much, you could extend the run. You've been doing some research and are keen to get some Araucanas for the blue eggs. You go to the market/an auction/a store. You buy a dozen mixed fertilised eggs, also an Araucana, and a couple of Frizzle Pekins because holy heck, have you seen them? SO CUTE.

4

The eggs hatch. Your garden and heart are full. Three of the chicks die, four of them are roosters, your cat gets another, one of your original girls becomes egg-bound and dies, one of your surviving chicks is a stunning Polish and she needs a friend so you decide to focus on rare breeds for a while and go shopping. If anyone asks how many chickens you have, you're genuinely confused, and settle on 'about half-a-dozen'.

5

You find yourself casually scanning properties for sale and fantasising about moving to the country, where you can have more land. And roosters.

6

You're gone, mate. →

THE FIRST RULE OF CHICKEN MATHS IS: YOU DO NOT TALK ABOUT CHICKEN MATHS.

Anyone who has chickens knows not to ask another person who has chickens how many chickens they have. It's the unwritten rule. When pushed, a chicken enthusiast will always round their numbers (way) down. You can pick a chook owner out of a line-up by enquiring how many chickens they have, then watching them wriggle like a worm on a hook. Listen for the qualifiers, the nervous laughter, the evasions. 'About thirty', 'I reckon forty-five', 'The council lets me have ten', 'The last time I counted, there were thirteen.' They are lying to your face and, in all likelihood, to themselves. They have many, many more chickens than they will own up to, and they do not want to think about it.

From the outside, this makes no sense. Logic dictates that chickens are finite, living entities. Surely countable. And yet, past single figures, no one can manage it. There are a number of contributing factors to this effect of chicken ownership. Chicken maths is a perfect storm of logic, emotion and philosophy. Here we go.

- Chicken maths kicks in as soon as you move past the purely transactional entry point of 'eggs'. Once you've bonded with your chooks, you want more chooks. They're entertaining, beautiful and calming. And you can get more! Access is the issue. Unlike cats or dogs, for which the socially acceptable ceiling is a couple of each, no one raises an eyebrow at a flock of chickens. Nature abhors a vacuum, and you will subconsciously come up with strategies and rationalisations to justify more chickens.
- There are hundreds of chook varieties, many of which are deliriously fancy. Once you start breed shopping, the 'collector' part of your brain kicks in. This is also the part of your brain that can always identify a need for another pair of shoes or another classic car.
- The variety, once you factor in colour, size and strain, is infinite. You will always find chickens that you don't have already, or that add to your collection, or are rare, or have desirable traits, or particularly beautiful feathers, or pink eggs, or . . . As with Pokémon, you've got to have them all. Once you give in to that, well, there's a certain degree of amused shame in knowing you're out of control.

- Chickens are very affordable. Even a fancy heritage breed chook will only set you back $60. That's a slab of beer. The only limit to the chooks in your life is your local council and self-control. We already know how self-control works out: part of chicken maths is plea-bargaining inside your own head, justifying why you *technically* have far fewer chickens than you actually do.
- More chickens are more fun than fewer chickens.
- Shrinking the size of your flock feels wrong.
- Not all chickens count. There's a reason we say, 'Don't count your chickens before they hatch.' Or even afterwards. Not all eggs hatch, not all hatched chicks survive, not all chicks that survive are female. If you've got a couple of young roos in your flock, they're not *actually* part of your flock because they'll need to be rehomed. Chicks aren't either: they might die or be roos. Depending on your philosophy, the following may or may not count: broody chickens, chooks who've stopped laying, chickens who haven't started laying, chickens you're giving away or selling, bantams or other poultry breeds such as turkeys or guineafowl.
- If you breed birds to show, you're attempting to create poultry perfection according to 'The Standards' and pick up some ribbons for Best of Breed. Or even save a breed. This requires a lot of breeding and hatching, and creates a great many 'am I not pretty enough?' birds who are surplus to requirements. Breeders never want to fully acknowledge how many excess chickens they've bred and are feeding and housing, and will need to cull, sell or give away. Or how much of their life and land have been given over to chickens. Hint: nearly all of it.
- If you decide to give a home to rescue chickens, your heart will break at how likely they are to die. They're bred that way. It is difficult to maintain a particular-sized flock because there are so many health variables. This means that a) you presume there's always room for more chickens, b) the size of the flock is never certain, and c) you want to save *all* battery chickens because their lives are so rotten.
- Watching a battery hen step on grass for the first time; watching them regain their feathers and their health; forming a trust-based relationship with a creature who has known nothing but fear and pain – that experience is *very* addictive. Many rescuers move to the country to have more space.
- Bantams, come on. They're tiny. They DO NOT count as full-sized chickens. You could fit three Rosecombs into one Australorp and still have room left for a quarter-sized pool table. If you have eight adult bantams and eight chicks, clearly that's at most only four chickens in total. This is probably not the argument I would use if the chicken guy from the council questioned why I have eight chickens[2] instead of the prescribed inner-city six. But I got rid of Percy as soon as I was contacted about the 'rooster issue', and the chicken guy was lovely and clearly on Team Chicken, so I reckon I'm okay.
- A chicken addict thinks this way all the time. There's always room for one more. It's like a problem drinker saying they only had two wines over dinner. The take-home, essentially, is that the place of chickens in your life expands to fill all available space. And then some.
- It is very difficult to explain all this to your partner if they're not as into chickens as you are. So you say things like, 'Last time I counted there were thirteen.'
- The world of chicken subcultures isn't particularly cohesive. Depending on the basis of your passion, there is much to disagree about. Activists are often vegans, and are appalled at, say, breeders, who cull a lot of their birds and, you know, eat them. Those who breed chickens for show despair at backyard crossbreed chooks, because there are historic heritage breeds that are endangered.
- Some people *still* breed roosters for fighting. Some believe rescuing chickens is a waste of time and resources, others that it's the only thing worth doing. Opinions and divisions are entrenched. But the thing that unites us all, no matter what our priorities are, is chicken maths and the deep and somewhat embarrassed understanding that we are, all of us, in the thrall of chooks. ●

2 Come on. They're BANTAMS.

CHICKEN CLUB

PETER WILSON

Principal, St Mark's Catholic Primary School

Works: Fawkner, Victoria

Titanja, Liam, Aranya and Andrew pictured with Pom Pom, a Silkie

'I've noticed the kids are right into food names. Like Chicken Nugget. And was the other one Roast? I'm not really keen on those.'

NUMBER OF CHICKENS:
Four (hens)

BREEDS:
Australorp and Silkie

NAMES:
Pom Pom, Bulbasaur, Shadow and Snowflake

Mr wilson
Every Day
Is A Good
DAY

Peter Wilson, the principal of St Mark's Catholic Primary School, founded by the Mercy Nuns in 1934, begins his typical work day at 7.10 am. He parks his car in the spot closest to the chicken coop, then goes in and mucks it out. 'I don't get the kids to do this. The kids do their jobs, but there are some jobs which have to be done thoroughly, otherwise we'd have piles of you-know-what everywhere.' He feeds the chooks, changes the water and leaves a couple of jobs for the members of the Chicken Club to do. 'They don't know I do this, by the way.'

Watching someone at their job, you get a sense very quickly if they're proficient. Peter is obviously an excellent principal. The kids, a multicultural surge of all ages, surround him in the playground, bringing their concerns and stories; he knows their names, where they're at, and speaks with kindness and respect. One boy, Liam, gives him a pencil drawing of a chicken, with 'Every day is a good day' written on it. Look, I've only got my own shoddy experiences to compare this with, but it seems like a pretty great school.

There is a sense that nothing goes unnoticed at St Mark's. Nothing is left to chance. Even the chooks are seamlessly integrated into the school's philosophy and teachings. 'The values of the Mercy Nuns are Inclusion, Welcome, Hospitality and Care. They're at the forefront of what we do with the chickens. We welcome them in, we care for them, we're hospitable, we meet their needs and make sure that they're happy. So, yes, very much part of our teachings.' Introducing the chickens has exceeded all of Peter's hopes for the school. He's smitten himself, with only one caveat: 'I do find them gluttonous.'

Peter had made chickens a goal when he became principal in 2015. 'I was conscious that we had some children who needed something during their breaks. You know, the ones who aren't your sporty kids, don't get involved in active games – they absolutely needed something. Recess can be very difficult and challenging.' His previous school had had chooks, and he'd observed that children would flock to the area – 'Pardon the pun.'

'It was helping their anxiety. They engaged with the chickens, and other children who were kindred spirits. And for children on the spectrum it was very beneficial. The connection with the chickens really calmed them, even when they had a meltdown. I wanted to bring that here.'

It's been a big success. Peter cites a boy who used the chooks to overcome his fear of birds. 'He was petrified. As soon as one moved he would just knock everyone out of the way to get out. But he set himself little goals, and he got to actually patting them, which is lovely.' Approximately 30 per cent of the school are in the Chicken Club, learning about looking after pets and responsible pet ownership by rotating through duties such as refreshing the water, cleaning out the run and collecting eggs according to an organised roster. In alignment with the school's social inclusion program, the eggs are provided to the Hope Cafe, a facility attached to the school which provides free meals on Thursday nights to those in need. St Mark's also provides produce from their gardens, and their social justice group, the Saints Club, cooks quiches for the cafe using the eggs. 'The kids from the club who are on that day physically take the eggs to the cafe, and they get to experience the notion of giving, and the joy that comes with that.' Mostly, though, they get to play with the chooks. 'They pick them up and put them under their arm, and have a lovely time with them.'

You can see it. The cuddles and the secret whispers. The care they take in holding the chook safely and kindly. The pleasure from giving them treats. Titanja Kalala – seven, pink bow, pink socks and tight braids – could hold Pom Pom the Silkie to eternity. 'They're my friends. I talk to them every time we have Chicken Club.' Aranya Ramawickrama, who shyly uses her inside voice, goes and sits with them when she's sad. 'They make me happy. And they tickle when I pet them.' Liam O'Dwyer, the chicken artist, aged nine, likes Shadow the 'lorp because she's 'kind of feisty'. Andrew Vimba, twelve, is a fan of Bulbasaur, because it was Andrew who suggested the name Bulbasaur (Yes! Result!). But after a moment's contemplation he lands on a more heartfelt sentiment: 'They bring us all together into one big family.' →

Peter beams. This is absolutely one of his values. There's only been one dark spot in the school's chicken adventure. In October 2017 two of the original school chickens, Fluffy and Nugget, were stolen during the school holidays (a crime that not only made the Channel Nine news, but surely has to be the epitome of bad karma). 'It was traumatic. We were trying to protect the children from the footage, but of course it came out in the media. We listened to how they were feeling, and supported them to feel safe. And that was a benefit, because it taught them that bad things can happen, but we've got ways we can recover.'

Part of the recovery was more chickens. Of course. 'I see a great benefit in them. I've got my own veggie patch at home, and when I'm closer to retirement, absolutely. Just to get out there and spend time with them: they're very therapeutic and calming. I could see that there's a parallel to prayer. It's a time when you can really focus in on your own thoughts.' ●

ENCHANTED KINGDOM

KAREN SAGER

Breeder of Pekins, weaver of spells, potato sorter, larrikin

Lives: Bundaberg, Queensland

Pictured with Peanut, a millefleur Pekin

'I don't usually sell my chickens. I end up giving them away. I let the chicken choose, and if it runs to you, if it chooses you, it's yours. But I'm fussy. If I don't like someone I won't sell to them. Even if there's a million chooks running around. No chicken for you!'

NUMBER OF CHICKENS:
Absolutely scads. Karen admits to 'more than fifteen'. Her son, Will, not yet grasping that chicken maths requires you to be evasive, runs off to count. He reports back with his serious face on: 'Sixty-one'.

BREEDS:
Araucana, Brahma and Pekin (white, black, porcelain and millefleur; hens and roosters). The Araucanas and Brahmas are Will's.

NAMES (STRAP YOURSELVES IN): BLACK PEKINS:
Rook, Raven, Hujen, Moonen and a rooster, Jebediah

WHITE PEKINS:
Stella, Nova, Astrid, Wednesday and a rooster, Rocket

PORCELAIN PEKINS:
Pia, Pearl, Pye (short for scrub python: she was bitten by one), Porsche, Summer and a rooster, Brian

MILLEFLEUR PEKINS:
Peanut, Winifred, Wendy, Lacey and a rooster, Bruce

ARAUCANAS:
Fraggle, Oracle, Super, Fugly, Blacky and Mini

BRAHMAS:
Hannah, Greystoke, Hetty, Ducky and Hey

'Babies and half-pints unnamed.'

OTHER PETS:
Three dogs, Lotti, a Belgian White Shepherd, Baby, a Kelpie cross, and Ruby Rose, a bitsa; eight sheep, Dot, Lilly, Rosie, Abbey, Winter, Bonnie, Cockie and Billy; a 250-kilogram pig, Piggy; native bees and butterflies

There's some kind of force field, a spell of concealment, around the six-acre Sager family property. It sits in the centre of an enchanted maze of increasingly unlikely roads that Apple Maps cannot penetrate. We're surrounded by trees on an unsealed road, outside of Bundaberg somewhere. Siri insists we turn up a path as faint as a paper cut between two lush fields. I ring Karen. She hoots. 'You're *where*? That's the middle of fucking nowhere. We're near as the crow flies,' she says, sending flurries of texts because her haven doesn't feature great phone coverage. She guides us in, with directions, jokes, emojis. It's an epic journey, and I get the sense of a veil parting, of being welcomed and initiated. Her final text: 'I have the cauldron warming and am chilling some beers.'

Not a joke, it turns out. There's a cauldron over an open fire – roasting dinner but, still, it conjures a definite fairytale air. As do the painted rocks, the purple streamer flag, the mosaics, the tiny bark fairy doors painted on trees, the fairies themselves arranged in nooks and crannies, and the twinkling LED lights. Much of this is tiny and hidden, a property-wide game Karen plays with the neighbouring children: a fantasy kingdom under our noses, just out of plain sight. 'What bliss! Sometimes I get cupcake cases and make tutus for the baby chickens.'

Karen, an improbable 45-year-old grandmother with lilac hair, star and butterfly tattoos, and a dress sense that borders on 'goth Pippi Longstocking', does what delights her. End of story. She wears a *lot* of purple and has a witch's hat for dress-ups. On a brief holiday from her job sorting potatoes on a local family farm, she's given herself a sensational set of glossy sculpted indigo nails. 'A little present for myself.' She's also addicted to breeding Pekins, which, no matter which way you slice it, is an impractical choice of fowl for a hard-scrabble limestone country property tilting at self-sufficiency.

'They're the little energies that never stop, around the yard. You look out and these little happy chickens, running after a grasshopper or something, look like they're picking up their frilly knickers and running. Like a cartoon. You always get a laugh. They're just my everything, really.'

Pekins aren't 'farm' chickens, by a long stretch. They're not a utility breed. They're a sweet, plump small ornamental bird, and essentially a carrier system for masses of glossy feathers. Karen's are comically broad in the beam. So much booty. They're a heart- not a head-choice, and the overall effect as they forage is surreal. Think an inverse *Wizard of Oz*, with no flying monkeys but a benign witch who sends out fat little feather balls to do her benevolent bidding.

'Most of my girls go to special-needs kids. One chicken was born blind in one eye, and she ended up with a little girl who was partially blind. You try and pair them up, you know what I mean? I have Pekins because they're gentle, beautiful, friendly darlings. I've bred for personality traits, for passive, loving natures, because that's as important as the purebred part. I just let nature do its thing. They're often still show quality; mostly they end up as pets. Loving and being loved is the most important thing. To me, giving someone a chicken, it's like giving love to someone that lasts as long as the lifetime of the chicken.'

Karen refuses to be defined by the expectations of others. She aims for what makes her happy, what is required to keep her dream functioning. It's too tricky, for example, to work out which Pekin eggs are fertilised, and a broody Pekin always gets to sit on its eggs, so the Sagers have Araucanas for the blue eggs, run by son Will. Husband Wayne, an arborist and ex-flower grower who's in charge of cooking as well as all of the building, knows the blue eggs are his. Problem solved. →

Will and Hannah, his Brahma.

Not a fan of the world beyond her borders, Karen prefers to not venture out much, carving out an oasis instead, a recycled wonderland that's ruled by kindness, a laugh, everyone pitching in and self-expression.

'Yeah, I prefer animals to most people. I don't like society or being out there. I go to the local shop to get bread and milk sometimes and come straight back. Self-preservation, I say. Because I can't cope. I hate it. There are too many people. Nastiness and horrible shit. It's all too fast, the rules are stupid, and I just don't like the world. It's bad. I have issues with it.'

The requirement for a haven came five years ago, after the family fled from their home in Ayr, unable to deal with the ice addicts and dealers living next door. Karen's fierce vision of a rollicking safe harbour is upheld by her loyal, equally eclectic family: Wayne ('If we had the soil I'd grow purple chrysanthemums tomorrow') and sons Mac-Len, sixteen ('I'm more of a plant person'), and fellow chicken enthusiast Will, thirteen ('Bit of a warning. I'm a hugger!'). Other family members visit often: her daughters Sage and Nissa, her grandkids Braxi and Tanner, her dad. Karen loves it when family arrive. 'Mum worries when Dad visits, because he doesn't go home for a few days. We end up having some large sessions.'

Karen's specialty is 'winging it every day on caffeine and weirdness'. She's been running her life on impulse and attitude since forever. 'I did my nail apprenticeship at fifteen. I said, "Fuck school and fuck the world," and started my own salon. I don't regret it at all. I don't regret not going to uni and dropping out like a bunch of my friends. Spending years pretending to be educated and losing ten years. Don't regret not doing that. I've loved it all.'

No matter where she's lived, Karen's always had chooks and space around her, although things have definitely escalated in Bundaberg. Besides chickens and family, there are sheep and a pig, a native beehive, butterflies, scorpions and snakes. Everything comes under Karen's care. Most of her menagerie find her; she's a beacon for runts and homeless creatures. She also goes to the dump and collects relics for the property. 'Old rusty things, things that have been loved and no one loves them anymore. Let them come here to die. Let them go back into the earth where they came from.'

If you need nurturing, find Karen and she will keep you safe. Are you an abandoned lamb? You'll be raised in the house. Pekin chicks whose mum was taken by a snake? You'll be raised in the house. Butterfly caterpillar? Taken into the house to cocoon safely in a jar and hatch. Runt of a pig who nearly died? You'll be raised in the house, where a photo will be taken of you watching *Babe* on television, and you will have a tutu made to fit, and you will come to weigh 250 kilograms. 'I don't know why I have a house. It ends up with crazy shit inside. I should just live in the shed.'

She flicks through her iPad, looking for the photo of a python after it ate five of her chickens, but gets lost in the flow of Chequered Swallowtail butterflies, native bees, flowers, goannas, turtles, tadpoles, scorpions, chickens, sheep and pig. She lands on a picture of a lamb dressed as a unicorn. 'Sorry, I get distracted.' She hoots. 'There's not many people on here, eh.'

'I can spend an entire day watching one flower. Watch it bloom. See how many bees visit it. People think I'm weird. But the fucks I give about what people think are low. Sometimes I think I'm lazy, that I should be doing something. But then I think . . . *nah*. I'll sit on the ridge and watch the scorpions. Lie in the hammock and watch the python overhead. Have a coffee and a cigarette with the chickens. They listen and they don't judge. They're very good to have conversations with. Yeah. I'm not part of society and I'm kind of proud of that.' ●

RESISTANCE IS FUTILE

EMILY HALLORAN

Chicken-vet-in-training

Lives: Adelaide Hills, South Australia

Pictured with Scarlett, a Faverolles

'What have I got in my bedroom? I've got a big chicken painting over the bed, and about three chicken figurines on my bedside table. There's a chicken basket that I put the eggs in, a shelf that's covered in chickens, and a dress and skirt and top with chickens on them. I've got chicken earrings. There's a chicken pen-holder, and another chicken painting. Oh, and there's a box under my bed full of little chicken things I've been given by friends.'

NUMBER OF CHICKENS:
Ten (seven hens, three roosters)

BREEDS:
Australorp, Barnevelder cross, Faverolles, Golden Laced Wyandotte, Golden Laced Wyandotte ISA Brown cross, ISA Brown Silkie cross

NAMES:
Midnight, Nancy, Scarlett, Luna, Florence, Ginger, Fluffy Feet and the roosters, Chickadee, Larry and Cornelius (named after the rooster on the Kellogg's Cornflakes box)

OTHER PETS:
Charli, a border collie, and Louie, a black and white cat

Emily Halloran, twenty-five and studying a Bachelor of Veterinary Bioscience at Adelaide University's Roseworthy Campus, experienced two 'Aha!' moments on her road to becoming a specialist chicken vet. The first was when her pet rooster Night, a Silkie cross bantam, almost died due to flystrike in his comb. 'I had to pull each maggot individually out of his head, and I did that for five weeks. It was totally disgusting. But he recovered, and I thought, "I've got this. If I can do this, I could work as a vet for chickens." Because it was, yeah, really gross.'

The second realisation was of the creeping variety: the vets she was taking her ailing chooks to weren't necessarily *across* chickens. And they certainly weren't taking her seriously. 'I've had lots of bad advice. I've taken chickens back and back when, really, I should have just put them down. The vets didn't know. They were guessing. And some of them can be so rude. They'll drop everything because the dog's sick, but if it's a chicken they're like, "Oh well". SO rude. My chickens mean the same as my dog and cat. They're *pets*. I love them. And there are a lot of us out there who do.'

'I'd love to have chickens at our wedding, to have photographs of us holding them.'

Pure of heart and determined (it is one thing to muse that being a chicken doctor might be nice, quite another to commit the years to training), Emily wants to be the caring vet who a) knows what she's doing, and b) won't shame or judge you for fronting up with a chicken. She gets it. She's fond of all animals, and they're drawn to her as if she were a Disney princess, but it's chickens that nestle in her heart of hearts.

'My sister, Annie, brought home three chicks from kindergarten to raise. Then I sort of took over, cared for them and fell for their personalities. I dedicated my afternoons to them instead of my homework. Then I kept adding and hatching more and more. I know it sounds silly, but they make me happy. If I'm feeling down I can go and talk to them. They're great to have sitting on your lap, to just stroke their feathers, and they fall asleep in the sun. I'll tell them what's stressing me out. I've got a real attachment. Every night when I lock them in, I give them each a hug and say goodnight by name.' →

It took a while for Emily to home in on her vocation. Keen to work with animals, she studied animal science with honours at Adelaide University, but it led her in the wrong direction. 'If I'd wanted to work with chickens it would have to be at, you know, a meat farm, and that's the last thing I'd want to do, to be honest.'

She's much more the caring kind. Fluffy Feet, named by her sister Annie, was hatched when Emily was in Year 8, which makes the Silkie cross twelve years old. Another died last year, aged fourteen, a gala lifespan for any chook, and over a hundred times the age a broiler chicken will reach. She's doing something right.

Emily's chooks have the run of a decent slice of woody hill out the back of her parents' property, which gives them plenty of dappled light and overhead cover, a must-have for good chicken mental health. Tending to the flock turned into valuable one-on-one time with her dad, Mark. They worked closely together when Emily was a child, and formed a solid bond. 'It was a good project to do with him. We built a coop, which turned into two coops, which turned into an extension. We're about due for another upgrade.'

Living on campus, Emily makes it home on the weekends, and finds it tough being away from her flock. 'Mum sends me photographs

sometimes, so I can see how they're going.' Her parents have also let her raise chicks in her bedroom ('What's a bit of poop between friends?'), and the walls of the house are hung with rather lovely linocuts of chickens, one of Emily's Year 12 art projects. 'My family understands that chickens are my special thing.' Her passion may be benign, but resistance is futile. Ask her fiancé.

'I've converted James to loving chickens. He didn't know much when we met, and wasn't excited by them. But then he got to know them. Now he's planning his coop, and saying things like, "When we move out together we'll have your chickens, of course, and then we'll have Rhode Island Reds and Wyandottes and three ISA Browns."' It sounds for the world like he's talking about their future children. In some ways, perhaps, he is. Smart man. Love Emily, love her chickens.

'I'd love to have chickens at our wedding, to have photographs of us holding them. It would, of course, be interesting holding a chicken with a white dress. We know what could happen.'

She smiles. Not put off in the slightest. ●

THE NAME GAME

HOW TO NAME YOUR CHICKENS

One of the great joys of chickens, no doubt, is naming them. It's an art, not a science. If you see chooks as an amorphous blob of 'they're just chickens', well, you're missing out – not only on experiencing chooks, nose to beak, as individuals, and on chicken cuddles, but also on the unfettered fun of what is essentially a low-stakes version of naming your children (nothing you pick will prevent your chook from getting into a good school).

So many options, depending on what story you want to tell. Punny or themed? A classic farmyard name that conjures bales of hay and chicks scratching for worms, or a witty pop culture reference? Princess Layer? Henny Penny? Ginger, Baby, Posh, Scary and Sporty? Are your chickens magnificent Kweens, or sensible aunties with shelf-like bosoms, or attention-seeking reality-TV stars, or Catholic saints? This is legitimate world-building, right at your fingertips.

As with everything to do with chickens, there's a light side as well as something deeper, darker and more subterranean at work. Naming is significant. Powerful, even. It's an important ritual, to look deep into your chicken and see who she or he is, and who you are reflected right back. If you have a chicken called Roast or Nugget, for example, you either have children to whom you gave unfiltered naming rights or you need to have a good look at yourself.

Naming a chicken changes the game. It endows meaning. It means claiming her (or him) as part of your family or tribe – paying enough attention to each chook to distinguish them and tell part of their story. It *gives* them a story. It changes how you treat them, and how much they trust you. It means, usually, that you're not going to kill them and eat them. For a chicken, that's an important distinction, and don't think they can't tell.

Chickens, unusually among domesticated critters, may be pets or livestock. Millions of them are culled or slaughtered every year, but some, through luck of hatching or rescue, are chosen to live it large in someone's back garden or hobby farm, rather than being intensively farmed for their meat or eggs. If you have chickens and you don't name them, you're keeping them at arm's length and, while that's a useful shield against engagement or feeling loss, well, let's just say you're not letting the sun shine in either. →

This is Joan Jett.

SO! HOW TO NAME YOUR CHICKENS?

HERE ARE A FEW THOUGHTS AND POINTERS

1

Wait for your chickens to grow up. I know, annoying. But naming chicks is utterly pointless. You don't know if they're boys or girls; some of them will die; their distinctive markings will change, and their qualities at that point are limited to 'adorable', 'peeping' and 'hungry'. It's important that the names reflect who the chooks are as adults, otherwise you'll find it difficult to remember who's who. Patience!

I've named a couple of chicks too early because they were sick, and it's effortful and a mixed bag. I didn't want them to die nameless, but it's tricky to come up with a respectful name for a ball of fluff of indeterminate gender.

For the little white chick violently attacked by her aunt Priscilla I went for Blondie. I thought if it was a girl she could be Debbie, and if a boy, Harry. Genius. Or, to put it another way, clearly overthought. She lived, and Blondie stuck, but to be honest I think she's an Elizabeth, as in Elizabeth I, Virgin Queen of England. All the scar tissue makes her neck feathers stick out weirdly like a ruff.

Val was a teeny 'blue' pompom close to death with a ghastly respiratory infection; I could have gone with Sneezy or Wheezy or Mucus Everywhere, but that's no legacy. I went for Val, short for Valiant, because she fought so hard to live. Some chooks give up. Those chooks are not Val.

2

Define your parameters. Decide whether you're going to name them individually, according to breed or as an entire flock. I once bought a pair of nervy ginger chickens at auction, before I understood that nervy chickens are a pain in the arse. They were something crossed with something else – sisters – and needed to be named as a pair. I ran through some options: Arsenic and Old Lace, Rum and Coke, Cinnamon and Nutmeg, Thelma and Louise, Cagney and Lacey, Bondage and Discipline, Drunk and Disorderly. I plumped for Laverne and Shirley in the end, which suited, given their unrelenting edge of hysteria.

The downside is, when one dies, you're stuck with half a duo, which complicates the origin story. You'll notice at Whiddon's aged care facility in Bathurst (see page 242), one of their chooks is called Pride. Which is a weird name for a chook, until you discover she originally had an offsider called Prejudice. But Prejudice died, which sounds like a philosophical win, and now Pride is top chook, so maybe she grew into it.

3

Pick a theme, which is to say, what's your passion? It's satisfying to combine flock-naming with nailing your colours to the mast. Also, a rare arena where the matriarchy can run riot uncontested. Kardashians? Feminists? Disney princesses? Literary heroines? Flowers? Dulux colour swatches? Song titles? Politicians? Fashion designers? Rock stars? TV characters? Historical figures? Cocktails? Sporting heroes? Artists? As yet I've not met any chickens named after *The Golden Girls*, and that's a gaping hole in the market right there.

There's something satisfying about a backyard with Virginia Woolf, Alice B. Toklas and Sylvia Pankhurst scratching around in it. Or Rose, Peony, Daisy and Triffid. Or Absinthe, Gin Fizz, Bubbles and Cask Red. Or Martina, Evonne, Ash and Venus. The choice of a particular theme can literally change the vibe of your garden. The pecking order automatically becomes more entertaining if you're seeing it through the competitive lens of women's football or characters from *Prisoner*, and your garden becomes more elegant if your chickens are named after the heiresses of *Downton Abbey*.

4

Be wary of naming your chickens after people you know. Not everyone sees it as a compliment, and it's not great if the chook dies. I named a sweet Australorp bantam Norah, for my ageing mum, and it was very awkward to deliver the depressing news a couple of months later that Norah had fallen off the perch. Not a good omen. Also, it is *very* tempting to name pushy or neurotic chickens after, say, a relative or friend who annoys you, or a bullying boss. Hilarious though it is, it's not worth the pain of a) having to explain it to the person's face, or b) the realisation that you're treating a chicken with less than love because you're projecting your dislike of their namesake. That chicken is *not* your mother or your geography teacher or your ex. Be nice. Choose names you're happy eggsplaining.

5

Go trad. Old lady names are always brilliant. They give off a nostalgic farmyard vibe that takes us right back to childhood and dignity and wholesomeness. You can't go wrong with Edith and Edna, Gladys and Daphne, May and June. You will definitely do more baking, and drink more tea.

6

Colours are always a solid option. It'll make it easier to distinguish between your gals and remember the names. If you get the vapours trying to commune with each chook and work out whether she's a Margaret Hatcher or a Fiona (go on) or a Lee Lin Chin, then start with their plumage and work back. What do they *look* like? What's the overall impression given by their colour, pattern, texture or shape? This way, you can come up with names that are memorable and recognisable and cute. Rusty, Raven, Honey, Peaches, Olive, Goldie, Creamy, Whitey, Blackie, Stripey, Red Rum or Brown Eye.

7

Puns. Of course. Chickens are simultaneously an ancient part of human culture and hilarious. They're playful animals and invite playful names. Amelia Egghart and Attila the Hen, Hen Solo and Mother Clucker. They're creative, and make us feel pretty good about ourselves for being such cunning linguists. Unfortunately, most of my good ones are for roosters, such as Bararck Obama or Russell Crow or Gregory Peck, and, like every other inner-city noob, I can't keep roosters. →

8

Pick a letter of the alphabet and stick to it. Great for larger flocks, or just as a way of narrowing your options to 'manageable'.

9

Build your flock from a variety of breeds. It will make them easier to name. The difficulty with a flock of ISA Browns or Hy-Lines or other utility chickens is that they look very, very similar. This makes it harder to distinguish them. I have a sneaking suspicion this works well for the companies that develop them. After all, you don't want to be seeing chickens as individuals if they're destined for a battery life, hey.

With more than two utilities, a leg-band helps, but if you're calling a chook 'Bluey' because it has a blue leg-band, are you really playing the personality game? When you're buying, put uniqueness near the top of your priority list. If you're a rescue-chicken enthusiast - and, if so, bless you and your work - then you'll be zoned in on personal quirks already. One of the joys of animal rescue is conferring dignity and individuality on creatures that have had it denied. Every chook deserves a story.

10

Finally, if you can, don't give your children unrestricted naming rights. Give them some kind of guidance and, I don't know, rules. Chickens can live for years, a decade, longer, and I've met chooks with names like KFC and Arghwomble. Random words are tricky to remember and have nothing to do with the chook. The joke names that imply they're going to be eaten have a short shelf life. Give your kids a few days to get to know their hens to prevent a 'blurt' name, and allow yourself a veto if they go for Poo Poo or Nando. On the other hand, the chook named Snot will remind you of your kids when they were little. Which is adorable, until the chook dies.

What can I say - it's complicated . . . ●

SAINT NIK

NIK ROUND

Advertising project director, tree-changer, heritage-breed enthusiast, chicken pimp

Lives: Trentham, Victoria

Pictured with a Rosecomb rooster

'We wanted chickens that would photograph well in our country cottage. Then it escalated.'

NUMBER OF CHICKENS:
Twenty to thirty

'When people ask me how many chickens I have, I always say, "How rude! How dare you ask!" No real chicken owner will ever tell you the truth. We always have many more than we like to admit.'

BREED:
Rosecomb bantam (black, blue, white and splash)

NAMES:
Nik is focused on the flock, not individuals. That said, the dominant black rooster is always called Rasputin. 'It's like the royals. When they become king they don't always get to keep their name.'

OTHER PETS:
A poodle and a spoodle, Bruno and Oliver, and bees

14

It took only three years for project director Nik Round to go from zero chickens to 'Best of Breed' award winner and Secretary of the Kyneton and District Poultry Club. 'I never enter into anything half-heartedly. I always want to be the best. Otherwise, what's the point?' he says, tweaking his favoured black rooster's tail into a photogenic spray. Rasputin the Third – framed to rustic perfection against Nik's bespoke leather apron and the idyllic, artful country garden – is ready for his close-up.

Nik, droll as all get out and full of charm, has form with fun hobbies and competitive animal breeding. As a schoolboy in Devon, England, he showed ferrets; in Australia he turned to horses and became a champion carriage driver. Australian champion in 2007, if you don't mind, and he represented Australia at the World Championship in Hungary. He has bees. But, even for an empire builder, the chicken thing took Nik by surprise.

'If I'm hungover the nicest thing to do is go in the garden, have a cup of tea, put a chicken in my lap and pet it.'

'I never thought of myself as a bird lover. Funny, beady-eyed, beaky things. I always saw myself as a bigger-animal person, you know, dogs and horses. And having chickens was just part of the whole backyard, country-cottage dream. But they're addictive, and now they're a major, defining part. Chickens grow on you. And, I tell you, if I'm hungover the nicest thing to do is go in the garden, have a cup of tea, put a chicken in my lap and pet it.

'They're mesmerising. Perfect little miniatures. I love seeing the flock moving together in the garden, almost in unison, the rooster in control. I lose hours that way. It's entertainment and relaxation; it's been proven to be good for your mental health. I call it chicken TV.'

Whether you're talking chickens or decor, Nik's attention to aesthetics is forensic. He has a great eye. His chickens are washed and oiled for competition, and the cottage he bought with his partner, Caz, is *styled*, baby. Duck-egg blue walls, blinding white enamel homeware, the latest edition chicken magazines splayed invitingly, and with wit, next to the toilet. Nik makes commercials and media campaigns ping, and has brought his skill set to Trentham. The result is precision rustic chic, which serves his need for beauty and order to function. It says much about Nik's poultry passion that the exquisite guest cottage is unusable: it's been requisitioned as a hatchery and is chockers with humming incubators. →

Mostly, though, Nik is having fun. He loves a chat, loves learning, loves passing his knowledge on, loves contributing, loves maximising his experience. His first Poultry Club win was an accident; he'd entered just for a kick and realised he was doing something right. His competitive spirit activated, and he's gone on to win several shows, including Australia's top Rosecomb award.

'That was much to the amusement of many. There were a lot of old boys there, and they said, "So, Nik, as you're new to Rosecombs, do you have anything to ask us?" And by this time I'd talked about chickens all day and, much as I love them, I was getting a bit bored and I was feeling cheeky, so I said, "Well. Seeing as I won the competition, maybe you boys would like to ask *me* something?" *That* went down like a cup of cold sick.'

Nik's chicken journey began nostalgically, a yearning for a country retreat that harked back to his parents' small holding in South West England. But he's not the type to be half-arsed, and in for a penny, in for a pound. His first birds were Silver Laced Wyandottes, which are almost stupidly attractive, but he's landed on Rosecombs for their classic beauty and their size. They're small, which means he can have more.

Although his birds are winning gongs, that's not the long game. Rosecombs are an ancient breed whose first flurry of popularity was in the fifteenth century, when King Richard III began raising them. Awake to the importance of breeding true, Nik has sworn to preserve and protect. The black Rosecomb has become popular in Australia, so Nik is focused on trying to revitalise the whites, splashes and blues.

'For a lot of people a chicken is an ISA Brown, and that's such a shame, because it's a forced breed designed to churn out eggs. The poor thing lays an egg a day, and that's just exhausting. There are many pure breeds of chickens that need to be looked after, some really rare, endangered bloodlines, and if people don't take care of them they won't be around. So, rather than having a backyard full of any old chickens, I encourage people to take an interest in one breed, and try and do something with it for the next generation. Pass it on.'

Nik walks the talk, keeping boxes of fertilised eggs marked 'splash' or 'blue' ready to dole out to anyone with a broody hen. 'Go on,' he says, 'don't make me scramble them.' He admits to palming his spare chickens off to friends and neighbours. Warm, supportive and as pushy as any matchmaker, he's responsible for a lot of Rosecombs finding homes in inner-city Melbourne.

'The most gratifying aspect is sharing the knowledge and chickens with other people. Giving chicks and eggs to friends. Passing on the chicken love. The *joy* it gives people, hatching out chicks in the summer, is fantastic. You can't underestimate the benefits. Chickens make you feel like you've got a home.'

Nik, who bought his Rosecombs from a local chap who went on to become his chicken mentor, is all about legacy and community. He still works and has a base in the city, but has committed entirely to local Trentham life through his show birds.

'I often say on Sundays that I'm off to do secret men's chicken business, and I head off and hang out in the poultry sheds. Chatting to old blokes in old sheds about birds. I'm so lucky. I interact with hundreds of people during the week at work – too many, sometimes – but on the weekend you can just have a good old chat and a cup of tea. The chicken club meets every week, and there's a nice extended community group. It's an opportunity to connect. It's a very lovely thing. There's a lot of retired men, and I enjoy spending time with people who are older than me, and learning from their wisdom. Mentors are so important. I'm almost fifty, but I'm considered the youngster, which is fun.'

Nik is an ideas man, never letting a chance go by. Why waste an opportunity? When he joined the Kyneton and District Poultry Club he noted that it needed to grow. The once creaky newsletters, social media and campaigns are now deadset swanky. And at poultry auctions and shows you will find Nik energetically front and centre, manning the kitchen, spruiking his chickens, coordinating any media, laying out chicken-themed napkins on long tables. →

'A lot of clubs all around the country were a vibrant and important part of the community, and many of them died out. Now there's a definite resurgence. There are a lot of tree-changers bringing new blood. The numbers at our local show doubled in 2016 from two hundred entries to four hundred. People want to connect back to the land, and chickens are the most simple way to do that.'

Nik's vision is to grow the success of the club, future-proofing it, opening the community to people who might once have felt excluded because they thought poultry shows were all farmers and old men.

'We've made an effort to make participation strips for all the kids. They've got to be included, and encouraged. It's very good for children to have the responsibility of looking after something, and even if they only spend a couple of years on it now, hopefully they'll remember, and when they're older they'll reinvest. It's about community legacy. Keep passing it on.' ●

CRAZY CHICKEN NERD

JENNIFER BIRKIN

Comedian, traffic controller, pop-culture obsessive

Lives: Adelaide, South Australia

Pictured with Willow, a Silkie

'All of my chickens are named after strong female sci-fi characters. There's been Xena Warrior Chicken, Princess Bubblegum and Phoenix from the *X-Men* – she was the first chicken to rise from the ashes after the first two died in a heatwave. I had a Buffy, until I discovered that Buffy was Spike and I had to take him back. Oh, and Uhura, from *Star Trek*. She was my first black chick.'

NUMBER OF CHICKENS:
Ten (hens)

BREEDS:
Hy-Line Brown, Golden Laced Wyandotte, Partridge Silkie, Speckled Sussex and White Leghorn cross

NAMES:
Cordelia and Willow (from *Buffy the Vampire Slayer*), Captain Janeway (from *Star Trek: Voyager*), Daria (from *Daria*), Lagertha (shieldmaiden from *Vikings*), BMO (asexual robot from the cartoon *Adventure Time*), Tash (Natasha Romanoff aka the Black Widow), Gamora (from *Strange Tales*), Fred (Winifred Burkle from *Angel*) and Oswin Oswald, for Dr Who's companion Clara Oswald and the password she used to hack the Great Intelligence

OTHER PETS:
Nah, too high maintenance

crazy chicken nerd

Jennifer Birkin, blunt smart-arse and chicken and pop-culture obsessive (she has 'Don't Panic' painted on the roof of her house), is performing her first solo comedy show, Crazy Chicken Nerd, in a dim bar underground lit with fairy lights and a single lamp.

'I've got 1051 photographs on my mobile phone, and 832 of them are of chickens,' says Jen, and the audience laughs, somewhere between impressed and shocked. 'They're revenge. People look offended when I say I don't want to see a photo of their children. They think I'm joking – because I do joke a lot – but I'm serious. No. I don't want to see it. It's a baby. They all look the same to me.'

It's the Adelaide Fringe Festival, and Biggie's basement, with its close black walls and graffiti-splashed toilets, has been pressed into service as a venue – alongside every other room in town bigger than a breadbox. The stage is not so much a stage as an eked-out corner of polished concrete floor. But a woman and her chicken don't need much room, and the audience, invested, lean in for Jen's talk about how her chickens have taken over, and to watch Willow peck around for scattered wheat. Jen's hand-reared Partridge Silkie, named for Buffy the Vampire Slayer's bestie, doesn't have any lines, but is very much the charismatic co-star. When she produces a substantial poo halfway through the show, everybody cheers, delighted.

'I've got 1051 photographs on my mobile phone, and 832 of them are of chickens.'

'Before the season I took the chickens to the bar a few times, to see how they'd go, to make sure they didn't mind the lighting or the music,' Jen says the next day, sitting in the backyard of her unpretentious old Adelaide house. She's framed by chooks and small building projects. 'They had to be happy. It was an experiment, and I was careful. Went around and told people I had a chicken and was about to take it out of the basket, because some people don't like flappy birds. But then I take her out and, of course, all anyone wants is to hold and pat the chicken.' Understandable. Jen enjoys patting the chickens herself. →

Chickens and comedy are Jen's blithe way of interacting with the world. And, to an extent, avoiding it. 'I can be an extrovert and an introvert. I can talk to anyone about anything – my grandmother said I was inoculated with a polygraph needle – but I can't stand stupid. People drive me nuts. I am almost obtusely independent.' She's ambivalent about many things. Other people. Relationships. Dogs and cats. 'They're too needy.' Joking is Jen's way – it leavens her candour – and the chooks supply the rest: entertainment and stories to tell, an occupation for her roving mind, undemanding company, eggs she can sell to the neighbours, and a tickling sense of self.

'Chickens are part of my identity. I'm known as the crazy chicken lady. Posting birthday pictures of chickens on everyone's Facebook page for the past four years probably hasn't helped. But chooks are perfect. They don't need me when I'm not there, but when I need them, here they are. I can go out at night and not worry. They're funny to talk about and better than a lot of people. They're not critical. They don't judge me and, yes, I have to deal with judgement. You're a woman who doesn't have kids, there's judgement. You walk down the street and a car will drive past and someone will yell out that you're fat. Thanks for that: I know, well done.'

Jen's got a lot of steel under the self-deprecation and bad chicken puns ('I'd like to thank the chick on sound'). She's complex and self-aware, and has her work-life balance enviably sorted. She's a trained nurse with a degree in genetics and zoology who reads science texts for fun, but these days chooses to work as a traffic controller. One of the people who hold up 'Stop' and 'Go' signs on roadworks, or guides pedestrians around potholes. 'It beats working for a living. It's easy. I don't have to think about it when I come home. I'd much rather think about chickens. And I do. A lot. They're good company.'

The chooks, and Jen's humour, come intertwined with a dark side. In Crazy Chicken Nerd she talks about her first chickens, the three she bought eight years ago for the eggs. Days later two of them were dead from one of Adelaide's notorious heatwaves. She buried them in the garden. Two days later Jen found her remaining chook scratching at the grave. 'Turns out she was eating the maggots. If I took one thing away, it's that the circle of life is fucking awful.'

Jen got into stand-up in 2005 after heckling at a gig; the comic asked when she was going to get up on stage herself. She made the Raw Comedy semifinal. She stopped performing after her ex-partner, Michael, committed suicide and her father died the year after. 'It got to the point where I didn't care if people laughed or not.' Crazy Chicken Nerd was the culmination of Jen's return to comedy, her chooks – in more than one way – lighting the path.

Having a bustling flock is part and parcel of who Jen is now. She takes pleasure from naming them, building their coops and feeders, keeping an eye on their antics and health, turning over the soil for them, plotting the next experiment. 'It's been good to have a focus outside of myself. I have to get up with them, and put them to bed at night. It's peaceful and relaxing to sit and just watch them wandering around. They make me happier.

'And I'll be quite happy to die in the backyard and have my chickens eat me. I can't take any of this seriously. Life's sad enough as it is. I'm not going to be miserable.' ●

The Honourable
Curtis Pitt

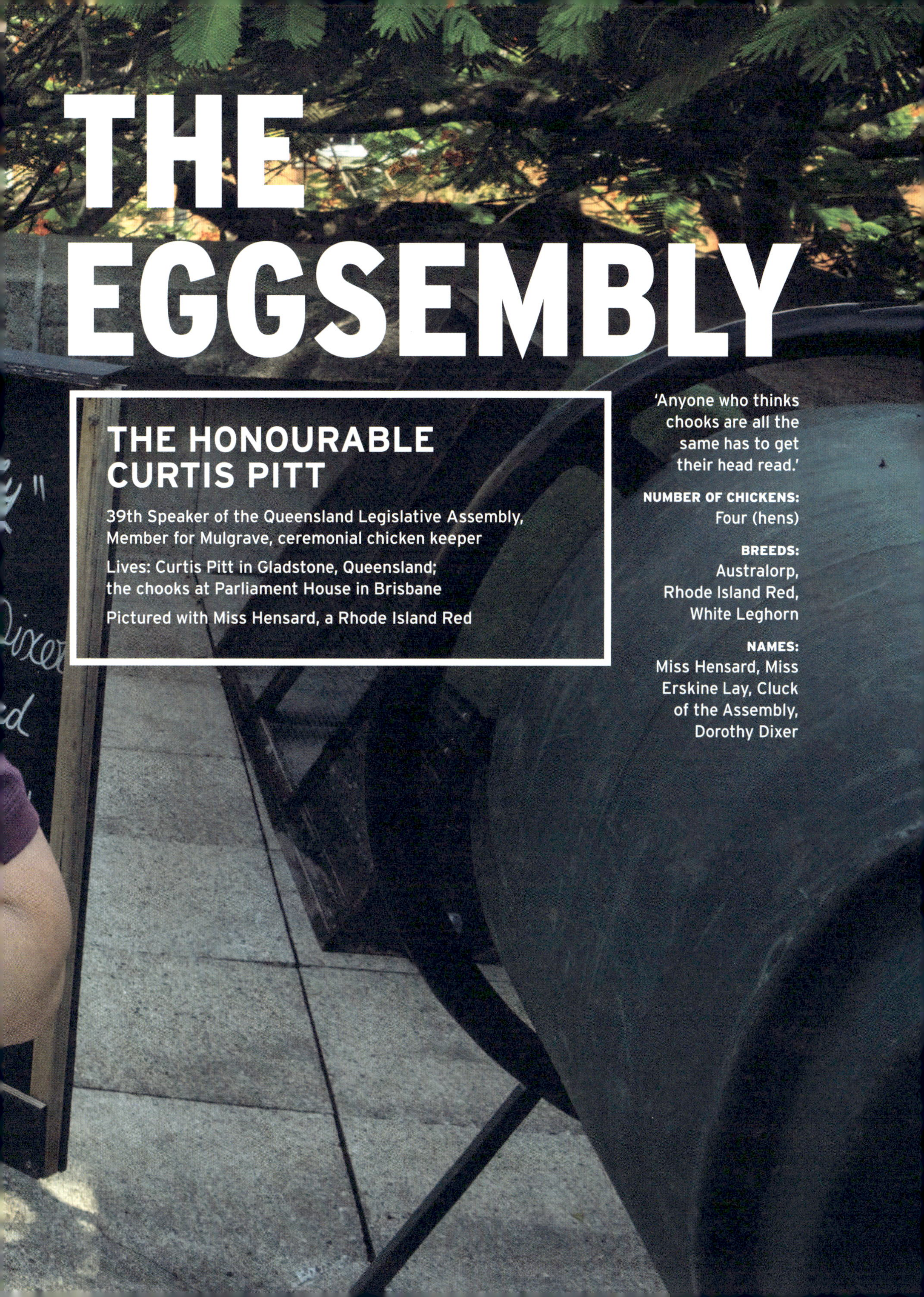

THE EGGSEMBLY

THE HONOURABLE CURTIS PITT

39th Speaker of the Queensland Legislative Assembly, Member for Mulgrave, ceremonial chicken keeper

Lives: Curtis Pitt in Gladstone, Queensland; the chooks at Parliament House in Brisbane

Pictured with Miss Hensard, a Rhode Island Red

'Anyone who thinks chooks are all the same has to get their head read.'

NUMBER OF CHICKENS:
Four (hens)

BREEDS:
Australorp, Rhode Island Red, White Leghorn

NAMES:
Miss Hensard, Miss Erskine Lay, Cluck of the Assembly, Dorothy Dixer

"Eggsembly"
Miss Hensard
Dorothy Dixer
Miss ErskineLay
Cluck of the Eggsembly

On 8 August 2017 – as set down in Hansard, the official record of business in the Queensland Parliament – proceedings were opened by the 38th Speaker of the Queensland Legislative Assembly and Member for Nicklin, the Honourable Peter Wellington, with a lighthearted welcome of four young hens. An interjector from the floor mentioned chicken curry, and Wellington clarified that the girls were off the menu, as valued additions to the House and the new vertical herb and edible flower garden. 'Another benefit to chooks,' he intoned, 'is that they have been found to have a calming influence on people and to lower stress levels. I accordingly invite members to visit the chooks, preferably before question time each day, so that perhaps we may see their calming influence and it may reduce my need to use standing order 253A.'

It is not recorded whether hilarity ensued, but given that order 253A refers to removing unruly MPs, and that former Queensland premier Joh Bjelke-Petersen referred to media briefings as 'feeding the chooks', it seems likely (and fitting) that there were a few hoots and hollers. Parliament is a dry, combative and process-driven work environment, and sitting MPs must grab any opportunity for levity and release. Chickens in the House – whether or not MPs take conscious advantage of their company – bring a certain *joie de vivre*. Wellington's address set the tone: the chooks occupy a curious intersection between political humbug and vaudeville, their presence letting the steam, ever so slightly, out of the parliamentary pressure cooker. Tended by gardening staff, they contribute eggs to the parliamentary kitchen, recycle scraps, provide a magnet for visitors, are great company for an outdoors lunch, and, hoo boy, do they provide a never-ending opportunity for humour. Puntastic, chicken-related humour.

A naming competition among staff threw up winners as eclectic and smart as you'd hope for: Miss Hensard, from 'Hansard'; Cluck of the Eggsembly, from 'clerk of the assembly'; Dorothy Dixer, after the entrenched practice of asking prearranged questions that allow parliamentarians to bang on about their favourite topic; and Miss Erskine Lay, after *Erskine May*, the authoritative guide to parliamentary practice and privilege published in the UK in 1844. Parliamentarians, clearly, need as much entertainment as they can get. →

The chooks' home was carved out of a pre-existing greenhouse in the riverside concourse of the parliamentary annex and, with whimsy in mind, named The Eggsembly, complete with official guide signs. It's kitsch, immaculate and fun, equipped with the mod cons you'd expect of a stockman's cottage, including a stove and curtains, window boxes with plants, a hanging kero lamp, and a framed print of an Australian outback scene that would sit well in a three-star motel. The girls' names are painted on wooden house signs hung outside on the front wall. It's an unexpected oasis of colour and charm in an otherwise officious environment.

The chickens were not acquired as a stunt; they were earnestly researched and obtained to recycle kitchen waste into fertiliser and eggs, to be an integral part of the sustainable vegetable and fruit garden, and, indeed, to bring a chill vibe. Still, at the open day on 11 August 2018, celebrating Queensland Parliament's 150th year, Miss Hensard, Cluck, Dorothy and Miss Lay were the runaway hits.

'There were thousands of people coming through, and this was the most popular exhibit. It's what people wanted to visit. The kids didn't want to do anything except come see the chooks,' says the Honourable Curtis Pitt, who took over from Wellington in 2018 as Speaker. In his ceremonial role as representative of the Parliament, Curtis has inherited responsibility for the chooks (who are immune to cabinet reshuffles and will outlast everyone). 'When my three children came down just after Christmas, they wanted to see the hens. I've had a limited amount of observation time myself. This is a busy place. Sadly, I think it's probably *too* busy sometimes. But, for a few people,

I think this garden near the chickens is a very nice place of solace.'

The Hon Speaker, to be fair, isn't out there of a morning collecting the eggs for the kitchen or raking the poo. Even if he were so inclined, the Queensland Parliament has a stiff delineation of duties, and a hierarchy, with two gardeners and a property director in charge of the chooks. What with security and OHS issues and finding keys and the Director of Property and Facility Services Stewart Johnson explaining their chicken husbandry practices, and the Honourable Speaker's Executive Officer George Hasanakos overseeing the handling of the official printed media statement, it takes four grown men in suits and lanyards to wrangle two of the chickens and juggle protocols within a prescribed twenty-minute interview window in the afternoon heat.

It's absurd, of course, and completely perfect. Because it's government, everything must be taken Very Seriously. The chickens each have official personnel sheets, like other employees, and the front desk staff have a binder with information documents used for school tours. The Hansard entry welcoming the chooks has been edited to read 'Hensard'. I ask for a copy and permission has to be sought from higher up. Is it allowed? It's official Parliament documentation! But it's chickens! Everyone's in on the joke, and no one's sure how seriously to take it.

'You have to put it in perspective,' Curtis says. 'It's not the biggest thing that's ever happened to Parliament.' Which the Hon Speaker knows to be true. In his career to date he's been Minister for Disability Services, Mental Health and Aboriginal and Torres Strait Islander partnerships; Minister for Employment and Industrial Relations; Minister for Trade and Investment; Treasurer and Deputy Government Whip. Standing there holding a chicken is . . . a weird thing for him to have to do. But still: important. The chickens, it transpires, light up the people who *do* get to spend time with them. The front desk staff illuminate at their mention. 'Oh, we *love* them,' says one. 'I go out and have lunch with them. It's relaxing. And the naming competition was brilliant, we had such fun with that.'

The security guard chuckles. 'We all love them here. And the tours. It's all they can talk about.' George, the Executive Officer, likes to come and spend ten minutes or so, hanging while they're scratching around. Stewart, the Director of Property Services, talks quietly and with satisfaction about the facilities provided for the girls (who are in meticulous good health), what provisions are made, how they're cared for. 'They run around and dig in the gardens, do what chickens do. And they come up to our office windows, because of the reflectiveness, and groom themselves in the mirror. We're very proud parents of the chickens. We love them. It's so nice having them around.'

So not a *big* change. But a change. And it might be worth tracking how many times standing order 253A has been used lately. ●

MOTHER HEN

EMMA GROSE

Peer support worker at Neami, a mental health recovery program for young people

Lives: Emma in Ferntree Gully, Victoria; the chooks at Neami in Noble Park, Victoria

Pictured with Nugget, an ISA Brown

'They take it seriously. Beyond what I hoped for and expected. It gets them up in the morning. Sometimes that's huge, particularly if they're on strong, anti-psychotic meds. One resident was spending days in bed, and the chickens were the only thing that got her up.'

NUMBER OF CHICKENS:
Five (hens)

BREEDS:
ISA Brown and Silkie

NAMES:
Svetlana, Saskia, Belafonte, Nugget and Gary

Maybe it's the soothing pastel walls, or the dependable blue couch, or the net curtains in the office with the box of tissues placed invitingly on the table, but it's impossible to escape the whiff of institution when you're in an institution. Even with the best of intentions, it's not a home. And that's tough when you're a young person with mental health issues: 'at risk', as they say in the documentation. When you don't *have* a home, or your home isn't safe, and you're pretty much convinced, anyway, that you're an actual scrape of dog poo on the shoe of life. Being in residence is fine. I mean, it's safe, and they're working on what ails you, but it's not your place, is it? It's not *home*.

It could be, though, if Neami peer support worker Emma Grose had her way. She knows how important 'home' is. As 'peer support', Emma has lived experience in mental health and recovery. Done her own time in a psychiatric ward. She knows what's needed. Lots of love and kindness, for one thing. 'They need a mum who will love them. A lot of them – if they have a family – the family's not great. I'm definitely not their mother, but I have a dream to have a farm, full of animals and full of young people who need someone to love them.'

The mental health recovery program at Neami's Noble Park residential facility takes up to ten young adults (aged sixteen to twenty-five), and the program lasts for a year. They arrive with diagnoses such as schizophrenia, autism, depression and anxiety. 'We have a lot of young women with borderline personality disorder, which is basically another name for trauma.' That was Emma's own diagnosis, back in the day. 'I was a chronic self-harmer,' she says matter-of-factly. 'Regularly attempted suicide.' But she's well into her own recovery now, and a large part of it (besides her place in the country, her partner, her dog and her kids) is her work at Neami. 'I have this great need to help everybody. I want to see people benefitting in the same way I did.'

There's scads of research available about the benefits of animal therapy, but three years back, when Emma was writing her proposal for getting chooks at Neami, she couldn't find anything specific about the benefits of therapy *chickens*. She knew it would work, but couldn't get any of her managers to take a punt. 'I came up against a world of *no*. No one wanted to hear my idea.' The facility has an ironclad 'no pets' policy, which was used to shut down conversation around obtaining the ex-battery hens that Emma wanted. 'I know the importance of animals. I wouldn't be here without animals. I ran a dog rescue organisation for six years, and that was my purpose. It saved me. Recent research found that 60 per cent of people who live with mental health issues included a pet in their inner circle of support; they're *important*. It's not practical for our young people to have their own pets. There'd be so many issues. But I wanted chickens as communal pets that everyone could be involved with. It had to be chickens! It was a community idea. I couldn't believe no one could hear what I wanted to say.'

Emma's luck changed when Neami's Victorian manager Glen Tobias visited the facility. She took a breath and pitched one more time. Glen has chickens. He said, 'Chickens are fantastic.' Glen was on board. The paperwork commenced.

It's now been two years since Emma and the then residents bought the flat-pack coop as a community project, and installed ex-battery ISA Browns Roast, Penny, Nugget, Gary, Helen and Hei Hei. There's been, as expected, turnover in both residents and chickens since then, but the results have been above and beyond Emma's hopes and expectations.

'Too many outcomes! It's been incredible. So many opportunities created that weren't there before. The chickens gave the residents a purpose, and formed them into a community. It gave them something to look forward to. They run it themselves.' The young people made up a chicken care roster, and take responsibility for feeding the chooks and putting them away. They get up and let them out early in the morning. 'That's at least five minutes of fresh air and sunshine they may not have got otherwise.'

One resident, after his housemate moved out, asked to change bedrooms. The chooks came up to the window of the other bedroom every morning, and he wanted to be near them. 'He moved the bed right up to the window.' One girl would go and sit on the ground in the garden, and the chooks would come up and gently peck her, as ex-battery hens do. 'She said, "This is the most grounding experience I've ever had. I feel like I'm here, in my body, for the first time." Now that's an outcome.' →

Some benefits have been unexpected, unpredictable: chickens as a pathway to conversation and communication. 'We had two new residents move in, and we'd got two new chickens, and the boys came to me and said, "We had to separate the chickens last night because they were attacking the new girls." Which was great anyway, that they did that themselves. And I said, "Yep, that's normal. It's basically just like throwing two new people into a house and going, oh, P.S. you live here now." And of course they were like, "Oh, *yeah*, we all know how that feels." Because that's what we do. If someone's housemate moves out, we literally plonk someone they don't know in the unit with them and go, "Oh, yeah, you're housemates now."'

The chickens achieved the seemingly impossible, turning Neami into more of a home. 'Rather than sitting in a room, which always has an institutional feel, I'll often say, "I just need to go and put the chickens away. Do you want to come with me and we'll have a chat?" And we'll sit out in the sun with the chickens for half an hour. It's relaxed and homey, and it's easier to talk when there's something else to focus on. I notice for a lot of the young men, our convos happen while we're cleaning out or feeding the chooks. It's perfectly acceptable to not make eye contact, when there's an animal present.

They're much more comfortable talking about something emotional when that pressure to make eye contact isn't there.'

There's more. So much more. The chooks contributing to staff wellbeing and self-care: everyone making excuses to take a break and go hang outside with the chickens. The eggs, obviously. Some residents learning how to be kind and gentle to animals for the first time. 'There's no question that if you've come from an abusive home, you've probably experienced animal abuse as well.' Oh. Of course they have. These young people have experienced so much. 'Look, this is why I want to be everybody's mum, because you just, oh my gosh, you just, your mind boggles. If we are disgusted about how people treat chickens, we should be having the same disgust towards people who treat their children the way they do. Because in so many ways it's the same.'

At the very heart of Emma's program, really, is one simple metaphor. She chose ex-battery hens so the young people in her care could experience taking a frightened, vulnerable critter from a powerless, abusive life, and watch it flourish when it's given love, kindness, security and attention. Pretty smart, eh? ●

CHOOKS AIN'T CHOOKS

HOW TO CHOOSE YOUR BREEDS

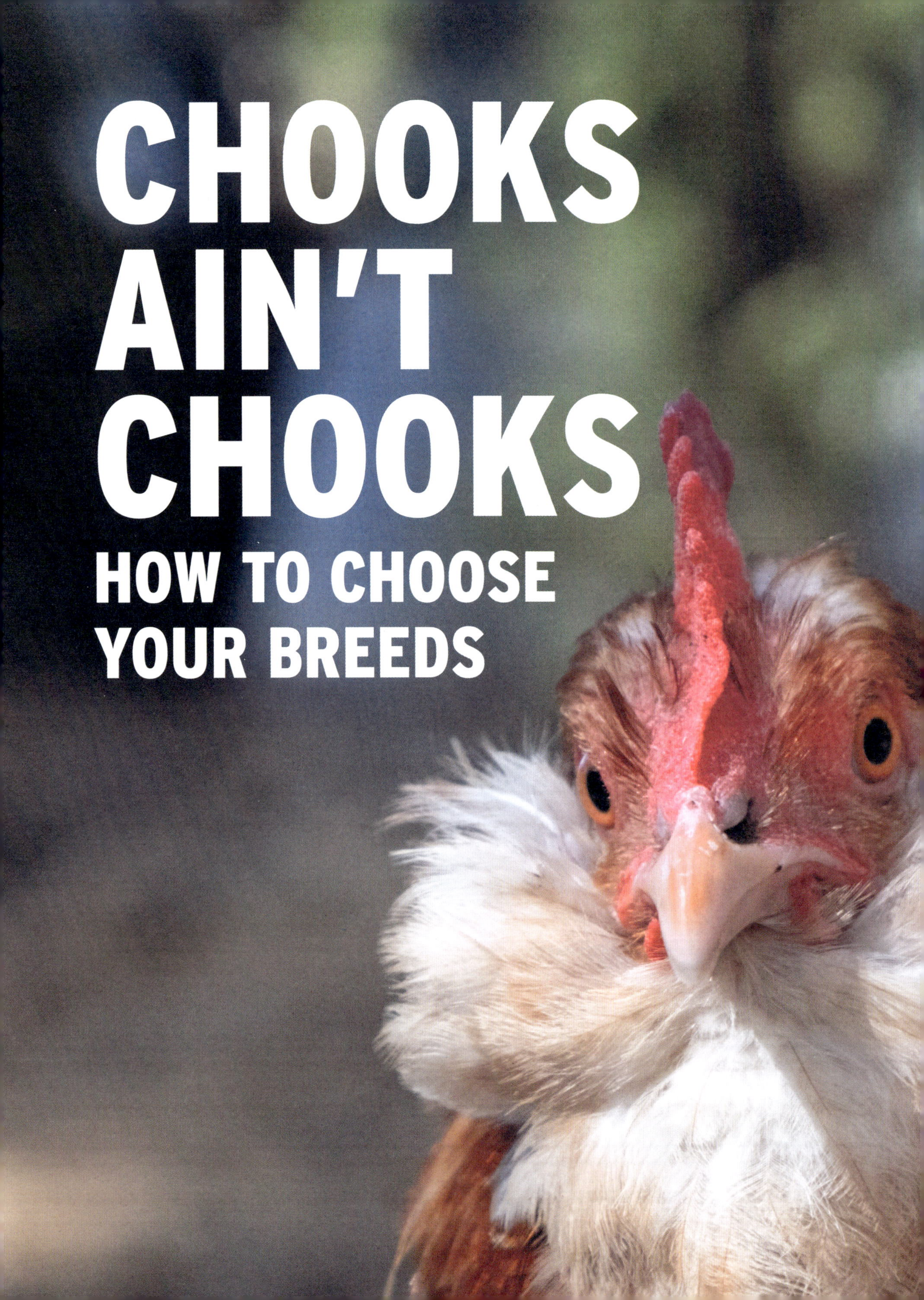

All chicken books, as deemed by law, must contain sections on, 'What are the different breeds of chicken?' And, 'What's the best breed to choose for your backyard?' And so forth. I enjoy them inasmuch as I like to flick through and see what they have to say about *my* chickens. In that sense they function much like a film or theatre review: justifying choices you've already made or are intending to make. There is, mysteriously, never an entry on Rosecomb bantams in backyard books, which is a huge failing, in my opinion. Just because they're flighty, lay infrequent small eggs and don't have enough meat on them to make a sandwich for a five-year-old - irrelevant. As if I'd eat Val!

I was a theatre and comedy reviewer for a very long time, and I know from dabbling in those dark arts that to bring depth to your recommendations you must be steeped in your subject. I am currently in the position of having a fair amount of knowledge about a handful of chook breeds and, at best, theoretical knowledge about a whole stack of others. Brahmas! They can grow very large![1] Buff Orpingtons! Ditto! Also: very fluffy bottoms! No one needs a chapter of that. Anything more I'd be getting from chicken books or the internet, and you can do that yourself. Hanging out with such legends as Alf Woods (see page 266) and Megg Miller (see page 176) has also taught me that to truly know a chicken, you have to run them yourself. I have what you would call 'niche' experience, to wit, my backyard has hosted Rosecombs, Silkies, Australorp bantams and Araucana crosses. The common theme you'll notice, if you're paying attention, is that my taste runs to bantams. Spoiler alert: I think you should get bantams. Unless you're contemplating permaculture, in which case that would be terrible advice.

I am seven years into my chicken journey. I've read widely, met many chickens and chicken people, and joined a startling number of backyard chicken, chicken rescue and exhibition poultry pages on Facebook. Here, under handy headlined sections, are my partisan hot takes. →

1 There's a video online of a Brahma rooster getting out of its coop on a farm in Kosovo, and it looks like a feathered Transformer unfolding. Just search for 'Brahma getting out of coop'. It's pretty much a pony.

YOU CAN'T GO TOO WRONG

The fabulous thing about chickens is they're incredibly varied and attractive. Push the boat out a little. There are some breeds that aren't considered 'entry level', because they've got genetic weaknesses from being bred to have fluorescent eyes and a fifteen-foot tail or whatever, but the chances of you stumbling on them are small unless you're ordering chickens on the dark web. Don't order chickens on the dark web: there's your first take-home.

Honestly? Within certain parameters I don't think it really matters what you get. (Controversial!) What matters is that you get some. You'll find out soon enough whether they suit your needs. Whatever you get you'll fall in love with. Every chicken will take you on a journey. The basic equation is: the kinder and more generous you are to *any* chicken, the more intimate the relationship. The chook's great gift and curse is that it will make do with whatever we give it, so if you go in with your heart open, an adventure awaits. You just need an entry point.

DECIDE WHY YOU WANT CHICKENS

It sounds self-evident, but break it down. As a pet? For the kids? For the eggs? For permaculture? For entertainment and mental health? To compete? This will help you choose. If you're wanting chickens to help turn your soil over and eat weeds, then you'll want an Australorp or Rhode Island Red. Something with strong legs that's going to get right in there. You don't want a bantam with furry feet that couldn't turn over a new leaf. If you want a pet for the kids, it's hard to go past a Silkie. They are hilarious, with a sweet nature, and easy to pick up and cuddle. Also Pekins. So fat and feathery! If you want one that will maybe fly up on your shoulder and loudly demand treats, a Rosecomb. But if you get your parameters sorted, there's no 'wrong'.

ENJOY THE PROCESS

I'm currently looking at buying a house – my first #freelancecreativelife – and I have to say, the level of research required and time spent Googling feel commensurate. Half the fun is in aspirational internet research: *Oooh, there's a coop made out of a 1950s caravan! Oh, you can get nappies! Check out that Phoenix rooster! Cochins come in eighty-three*[2] *colours and one of them matches my hat! Lawks, that Polish looks like it just won* RuPaul's Drag Race*!* For the full experience, let yourself go a bit. Even if you're all about the eggs, there are a lot of options. Chickens, if you let them, can be half livestock, half pet and half portal to a mystical dimension. I know that's three halves. Chicken maths. Mic drop.

GO HERITAGE

This is my big takeaway. There are so many cool and unique breeds of chicken that are under threat. The Rosecomb, for example. We've been so swamped by utility hybrids that we've come to presume that all chickens *are* ISA Browns. That a chicken, by default, is a brown bird with no identifying features. This is end-stage capitalism at its finest, destroying diversity within a species for profit and efficiency. How would you like it if there were only one breed of dog, and they all died aged two? There are a couple of hundred breeds of heritage chicken, fifty of which are recognised in Australia for show purposes and included in The Standards. You will be helping to maintain pure bloodlines and, in a small way, even helping the ISA Brown by reducing demand for that poor beleaguered chook. Also, heritage breeds live longer, lay longer and are much more interesting to look at.

MAYBE DON'T GET AN ISA BROWN

There's a lot of advice that you should start out with ISA Browns, or Hy-Lines, or another utility chook, as they're considered an 'entry-level' chicken. The more I know about them, the less I'm sure about that. Yes, they're loving and good-natured and lay lots of eggs. But so are heritage breeds. And therein lies the issue. Buying a hybrid is supporting the industrialisation of poultry. It does nothing to further the world of poultry. ISAs aren't so much a breed of chicken as they are an actual brand, like Coca-Cola. ISA stands for Institut de Sélection Animale, after the French company that designed them, in 1978, to produce lots of eggs in a battery environment. The 'brown', I guess, is self-evident. Go to the institute's website, it's chilling. The lead photograph is of two science guys in a laboratory, and the blurb talks about how 'ISA's products achieve exceptional results in efficiency and production'. If you hatch from their 'products', you don't

2 Not eighty-three. But loads.

get more ISAs, they don't breed true. This is because ISAs are an end product, made from a carefully guarded secret recipe of genetic ingredients.

It's the 'egg a day' part which is the problem. They're bred to lay at the expense of everything, including their own bodies. They have a short laying life because their reproductive system has been turbo-charged, which means eggs for a couple of years and then they're spent; they stop laying. Humans did not create them to live long, and they don't. Two to four years. They're created to be slaughtered at eighteen months old. The same way that meat chicken hybrids have been created to be slaughtered at seven weeks old. You can give these chickens as lovely a life as you like, but they're programmed to self-destruct, and the best thing we can do for them, actually, is stop creating demand for them.

Layer hybrids are destined for reproductive issues and get egg-bound, reproductive tumours and other sad ghastly stuff. The back end of a hybrid's life is not pretty, metaphorically or literally. They get ratty from using all of their available protein on eggs rather than feathers. Their big advantage is that you're not blinded by choice - I'll take these three brown ones, thanks! - but there are a great many intriguing chickens out there, so why not lift the bar a little? A heritage breed will lay almost as well[3], for a longer period, and will live significantly longer too.

If you think you want a 'layer', understand that all lady chickens *are* that, by definition.

MAKE SURE YOU'RE SURE IF YOU'RE THINKING RESCUE CHICKENS

I genuinely applaud those who take on rescue chickens. It's important and valuable work. But it is *work*. There is so much joy to be had from giving a traumatised bird a new life. It's not to be taken on lightly, though. They need so much care, love and resources. You probably won't be getting eggs, and their health issues mean they can't easily run with a regular flock. I guess what I'm saying is that rescue chickens aren't entry level.

GET A VARIETY OF HERITAGE CHOOKS

This speaks to the hybrid issue again. Chickens are much more fun if you can easily tell them apart. It personalises them. Get different colours or patterns in the same breed, or different breeds altogether. It's wise to get at least a couple of each breed, though, as they do tend to hang with their own kind. They won't necessarily buddy up across breeds. A helpful strategy is to keep an eye on chook size. In my second flock I acquired a couple of Araucana crosses in the hope of getting blue eggs. They were much larger than the two young Australorp bantams I'd bought, and the Araucanas turned into total hench-chickens who delighted in monstering the little 'lorps. Because they could. Also, I did not get blue eggs. Would not do again.

DECIDE ON BANTAM OR FULL SIZE

Full-size chooks can, if you're not careful, trash your garden. That may be fine, or even what you're after, but I was a bit shocked at how quickly 'the twins' (I couldn't tell the Araucanas apart) destroyed the joint just by being chickens. Bantams are much lighter on the garden and, in fact, offer you the opportunity to have one. Cards on the table: I think, for lightweights like me who are intoxicated with the prettiness of the different breeds and their company, rather than What They Do For The Garden, bantams are perfect. They're so entertaining, in all their variety and foofiness. Their eggs are small, rich and delicious. And their poo is not overwhelming (big chickens do produce a lot of black gold).

WHERE SHOULD I BUY FROM?

Your local poultry auction, 100 per cent. There is nothing like the atmosphere. The sheer variety of chickens on display will blow your hair back; you'll meet some absolutely fascinating characters, and learn a truckload about chickens from a shed full of obsessives. It's also cheaper, and you can make a day of it. Even if you're not buying, it's a terrific opportunity to window-shop. The chooks are also likely to be healthy, as poultry clubs are stringent. Other solid options are farms, friends and chook pages online. Make sure you quarantine any new gals for a couple of weeks from an existing flock, to make sure they're not carrying any disease. →

3 They will, however, take a break/slow down during winter. But we all need a holiday!

BUY A HAPPY-LOOKING CHOOK
If you're at an auction or market, go for the bright-eyed, curious, brave chickens. The ones at the front of their enclosure, checking out what's up, or even busting to get out. A chook that's cowering at the back of its cage isn't 'shy', it's anxious and probably scared of humans. Also maybe sick. I made the mistake of buying a gorgeous couple of crossbreed ginger chooks at a silent auction. They were quiet and huddled at the back and I was sorry for them. Shirley sort of turned out okay, but Laverne was a massive pain in the neck because no matter what I did she ran away. Tasty treats? NO WAY, YOU'RE TRYING TO KILL ME. It's exhausting and boring. Go for outgoing girls; they'll give you a lot more joy and a fighting chance of building a relationship.

SPEAKING OF OUTGOING GIRLS: ROSECOMBS
If you're after a decorative addition to your backyard, and a small inquisitive chook with dark eyes that can fly rather well, look no further! The Rosecomb Bantam Association of America describes them as 'the beautiful little aristocrat of the bantam world', and they are proud, fancy, saucy little chickens that are considered 'ornamental' rather than 'useful' (they're not big egg producers). They have a large 'rose' comb, with striking white earlobes and a *biiig* personality. Eggstroverts. I like my chickens to be companions and entertainment, so they work for me.

MY SECRET FAVOURITE BREED IS: AUSTRALORP BANTAMS
They were my first foray into chooks, and I did love them. Good egg layers, gorgeous green-sheened black feathers, black liquid eyes and outgoing natures. (Don't tell Val.)

AND, FINALLY, EVERYONE SHOULD HAVE A COUPLE OF SILKIES
When most people think 'bantam', they think 'Silkie', and such is the ubiquity of this unlikely explosion of fluff that it's easy to understand why. A favourite with everybody, the Silkie pops up all over due to its kid- and cuddle-friendly personality, striking Instagrammable looks, and the feel of a rabbit or kitten. They're an ornamental bird, and do a good impersonation of a seventies fashion model dressed for a 'this season's fur' snow shoot. Or, you know, a muppet. They manage the neat trick of being both freely available and genuinely extraordinary. Hailing from China in the twelfth century, they were mentioned in dispatches by Marco Polo and exist to be cosseted. Up close they're alien, with turquoise earlobes, five toes, furry feet and black or blue skin, flesh and bones. I'll say that again: black or blue skin, and black *bones*. Wild.

Their 'fur' is due to a mutation in their feathers, which means they lack functioning barbicels. Barbicels are the 'hooks' on feather fibres which link together and give feathers shape and form. Silkies are essentially big balls of flightless fluff that go broody at the drop of a hat. They're the devoted mothers of the chicken world. Show them a golf ball and they will absolutely try to hatch it. Their eggs are small and delicious, and they lay reasonably well. Their MO is to lay a clutch of eggs – say, a couple of weeks' worth – and then go off the lay to sit on them. The broodiness can get old pretty quick if you *don't* want them to hatch eggs. Just as well they're adorable. One of mine, Phyllis, technically white, likes dirt baths and dunking her 'fro in mud so that it dries into spikes and allows her to see properly. Every other Silkie I've met or seen is pristine and coiffed. Phyllis is filthy twenty-four seven, and unrepentant.

There are plenty of other cute, sweet-natured heritage chickens, but a couple of Silkies in the mix means you'll always have the 'Aw' factor. Also, someone to stand on your feet when you're in the garden. ●

CHENIUS

MAX COSGROVE

Chicken entrepreneur, fundraiser for breast cancer research, inventor of the Cheanie, the Chumper, the Chickini, the Chickinator, the Chandbag and the Chanklet. Ideas man, schoolboy, son.

Lives: Machine Creek (population 120), Queensland

Belinda and Max pictured with Captain, a Rhode Island Red rooster

'I started when I was seven, with two chickens. And then it just kept going and going. This breed, that breed. They're addictive.'

NUMBER OF CHICKENS: About a hundred at any one time (hens and roosters)

BREEDS: Australorp, Blue Laced Red Wyandotte, Coronation Sussex, Frizzle, Lavender Sussex, Pekin, Plymouth Rock, Polish, Rhode Island Red, Silkie, Silver Laced Wyandotte

NAMES (OF FAVOURITES): Captain, Favourite Boy, Princess II and Peaches II, Lady Pants and Smoky

OTHER PETS: A rottweiler, Rocky; two Australian cattle dogs, Izzy and Rippa; and four Brahman cows, Cuddles, Lucinda, Lagoon and Master

Life is
BETTER
On The
Max
May 2019

Max Cosgrove has startling blue-green eyes with a distant gaze. The kind of gaze you get, probably, when the back deck of the family cattle farm overlooks an endless rolling plain. When you're an only child. When your mum has had a scary battle with breast cancer and you've had to do a whole lot of stepping up. The rest of Max is sturdy, cheery and accommodating: a young fellow of few words, and Australia's most unlikely entrepreneur. A twelve-year-old with a runaway line of chicken fashions and ankle identity bracelets. Or, as they're known in Max's product universe, 'Chanklets'. They're also the first product in his planned range of chicken jewellery. Sorry: 'Choollery'.

Ready with a thumbs-up and an eager smile for a customer or the media, wearing a wide-brimmed hat and buttoned blue shirt with 'Max's Chickens, EggSellent to the MAX' embroidered over his heart, Max has been self-branding for years. He's appeared on Channel 7's *Little Big Shots* and *Australia's Got Talent*, been approached by an American talent show to train a chicken to ride a skateboard, and is regularly featured online in news stories written for Queensland's *ABC Rural* and *The Observer*. He's been interviewed by Alan Jones on 2GB. He's a surprising one-boy generator of quirky, feel-good, chicken-forward stories that raise a smile. To contemplate Max's achievements is to stare into the abyss of your own comparative laziness and inadequacy. What did you do today, champ? I guarantee that Max put his head down and did more.

Max spends four hours a day with his flocks of purebred chooks. An hour in the morning, feeding and watering, then an afternoon of cleaning, checking the incubator, and letting them all out to free range. There's school, of course – where he's house captain of the softball team – but also his four cows (Cuddles, Lucinda, Lagoon and Master) to take care of, and fences to be checked. Then there's the shop he's setting up on the property to sell his special chicken-feed mix and hand-riveted chicken feeders, his branded Chook Poo Brew fertiliser, the chicken beanies and jumpers knitted by his nana and great-aunt, roo ties for the roosters, chicken fascinators, and little chick coin purses. He puts out his own calendar. He sells extendable mango pickers and organic mangos. He has a stall at the Mount Larcom farmers' market, sells chooks through the Max's Chickens Facebook page (more than 11,000 subscribers), and has won ribbons for Handling and Best of Breed at the Mount Larcom Show. 'We have to call him in at the end of the day. He'd be out there all night,' says his mum, Belinda.

Max also spends ten hours a week sewing his 'Chickini' range – bikinis for chickens – which he thought of after his mum underwent a double mastectomy. The 'Cheanies' were inspired by her hair loss. He donates a dollar from every chook-clothing sale to the Breast Cancer Foundation. Over $1000 so far. 'When Mum had chemo. That's when I started. I like . . . I like the idea of giving. It's hard to go through breast cancer as a family, so, yeah, it's good to support.'

Belinda, currently on health leave, stitched back together from steel and heart, struggles to articulate how impressed she is. How touched. 'I can't really find the right word for it, to be honest, do you know? The *Courier Mail* man just said, "You must be proud," and I said, "Very proud." But I don't think "proud" is the right word. Like, what's a better word than proud? Awe, I guess. I'm in awe of him.'

Max was born with the hustle. His first venture, aged five, was a busking expedition out the front of the farm (on a road with zero drive-by traffic), followed by an equally doomed roadside stall selling his toys. When he was seven a teacher gave him a couple of hens, and he toyed with selling eggs before realising the chickens themselves would be more profitable. He saved up for a 'quad' (a rooster and three hens), borrowed money from his parents to buy an incubator, and read books from the local library. When he had the inspiration for Chickinis, he asked for a sewing machine and took lessons.

'My husband often says Max only ever needs to be shown something once,' says Belinda. 'Then he can do it.' She doesn't so much push Max as watch him go. His dad, Laurie, is a fitter and turner with Rio Tinto, and neither of them anticipated the 'Chempire' (chicken empire). 'We thought the chickens would be like a little hobby. We had the space, and we thought, "Oh yeah, it'll be a bit of fun to watch the chickens hatch."' →

Within two years Max had ten pure breeds of chicken (about a hundred birds), a high profile and a loyal clientele. Tickled by his youth and enterprise, people drive for hours to buy his chickens and merch. He paid back the incubator loan and bought his own quad bike and skateboard, several head of cattle and an affectionate rottweiler called Rocky. He's saving to buy the house down the road, and has his own debit card for supplies. He puts money into upgrading the chook pens and buying the chickens treats. He's also bought his mum jewellery, including a silver 'tree of life' necklace. 'That was right on when I was diagnosed,' she says. 'He's always been kind.'

Chooks are business, but they mean a lot more. It doesn't take Max three hours to tend them of an arvo. He just sits with them. Chatting. Hangs out. Has a cuddle with his favourite, Captain, a large and placid Rhode Island Red rooster who comes up well in a roo tie. Sometimes Max does his maths homework in one of the pens. 'They've got different personalities. I like the calm ones. I can pet and hold them.' He had two loved Silkies, Peaches and Princess, who followed him around (and modelled the chickinis and chumpers), but a fox came. Max has picked successors: their daughters, Princess II and Peaches II. 'They're very soft.'

He cares for his chickens. He cares full stop. 'When I was sick,' says Belinda, 'he treated me like his baby. I was too crook to do anything, and he'd be out there getting himself ready, making his own lunches, making me something before going to school. I'd watch him get on the bus then I'd sleep all day, set my alarm to watch him get off the bus, and he'd throw down his bag, get me a water, a cup of tea, look after me. He ran me baths, because I ached so much. It must have been hard, but he didn't complain, not once.'

Max doesn't even complain to the chickens. 'No way. I just sit with them. It's quiet. Because they don't talk back.'

Belinda laughs. 'You're funny. You go down there for peace and quiet? That would be true. Time out from looking after me.'

Max grins, seeking out his mother's gaze. 'Yes. It was.'

There's a slight kerfuffle with Rocky, the big galoot of a Rottie, who knocks into a chair; Captain, taking the opportunity, flaps down from the bench he was modelling on and sprints the 100 metres towards the pens, bow tie flapping behind in. Max takes off in hilarious, futile pursuit, looking less like an entrepreneur with a chicken-accessory empire, and, momentarily, exactly like what he is: a twelve-year-old boy chasing a chook. ●

Max's Chickens
EggSellent to the MAX
Max
RB SELLARS

Dress up your favourite girl in a Chicken Bikini!
The CHICKINI

THE GOOD LIFE

CASSIE MULKERIN

Hippy, romantic, proud check-out assistant at Pasadena Foodland – 'The finest supermarket in Adelaide! It has a restaurant.'

Lives: Clovelly Park, South Australia

Pictured with Pepsi, a crossbreed

'"If an egg is broken by an outside force, life ends. If broken by an inside force, life begins. Great things always begin from the inside." I love that saying.'

NUMBER OF CHICKENS: Ten (nine hens, one rooster)

BREEDS: Barnevelder, Lavender Sussex, Polish Wyandotte cross

NAMES: Pepsi, Max, Lyra, Lovely, Martha, Sunflower, Sophie, Lucille, Sylvie and Bellarose

OTHER PETS: Audrey, a Hungarian Vizla; Olly, a rescue cat; and Esther, Coco and Otto, mini lop rabbits

Of a warm, welcoming, early evening in Adelaide, you'll find Cassie Mulkerin with a glass of wine, out in the back garden with her chickens, paying close attention to their behaviour. She's there in the morning, too, taking respite from family duties with a coffee, hanging near the chooks' cool, grape-laden run, working out what makes her flock tick. She'll sit in their shed, and tell Max how handsome he is, and urge the hens to go broody. 'Come on, I want you to make babies for me.' A spiritual friend, Marnie, has advised that chickens respond to the vibrations in music, and sometimes she dials up YouTube on her phone for them, playing selections from soYmartino's Fingerstyle Guitar School. 'They do seem to gather round and pay attention.' She's also considering bringing portable speakers out and playing a bit of Beth Orton. Some Xavier Rudd, perhaps. 'Definitely something folky.'

Cassie feels deeply connected to animals, and is edging towards becoming a natural animal healer. 'I spend a lot of time watching the chickens, noticing what they choose in their own environment, how they look after themselves. I'm just watching and learning.' She points to her rooster, Max. 'See that? He's telling her there's food there. He thinks it's a good place to lay, and he's been letting her know for days. He's like, "Come on! Over here! I've got all this food for you!"' The object of Max's frantic attention, Pepsi, is having none of it. 'She knows what she's doing. She doesn't need to be shown.'

The archetypal sweet blonde hippy chick, with a ready smile, a sun and wave tattoo on her wrist, a green shift dress, ugg boots and zero interest in being constrained by anything, such as other people's values or a bra, Cassie is living her chosen simple life in suburban Adelaide. 'We call it our backyard farm.' It's a faintly dishevelled, gloriously homey set-up. Old-school. Cassie's got it sorted. Her chickens, her childhood sweetheart and husband, Ben, bunnies, a dog, a cat and their three kids, Hannah, Noah and Zoe, who run wild after school through the vegetable gardens. 'They love it – the chickens get them outside. Zoe will sit there for hours and let chicks jump all over her.' Cassie worked for ten years as a nurse, but the shift work was too hard with three children. Now she works check-out at Adelaide's schmanciest supermarket, which gives her the flexibility she needs with a tribe aged seven, nine and fourteen. Ben works for the government in public transport, and is passionate about computer gaming and growing vegetables. They met in Year 12, married at twenty-two, and fifteen years later are still happy as clams.

'We're trying to not focus so much on money, which the world seems to do these days. We want to appreciate other things in life, such as having a beautiful family and enjoying that love and bond. People strive to have an amazing show home, but we could never live like that. We love our humble little home, even though the carpet's from the seventies. Who cares? We want the kids to be able to focus on the things they enjoy, and strive to do what they want, and not what they're told or expected to do, you know?'

There's steel behind this laidback philosophy. The edge that comes from having been diverted from your own path. It was deadening having an instinct for animals yet growing up without them. 'Dad didn't like pets. I'm not sure why. Maybe they thought a pet was just another expense. Too much looking after. I'm one of five girls. The excuse was always, "We don't have a gate and the dog would run out into the road."' She shrugs. Gives Pepsi a squeeze. 'I think I really missed out. It was the one thing that was embedded in me somewhere. The need for an animal.'

Cassie wanted to be a veterinary nurse but, concerned it wouldn't pay well, her parents channelled her into regular, human, nursing. 'I just had to let the dream go, to keep them happy.' Obviously on a roll, her parents weren't that keen on Ben either, but Cassie dug her heels in. 'I don't know that they even really know him now. He's such a good man. So sweet, so caring and nurturing. We're lucky, we grew together, stayed on the path together.' Cassie's certain about Ben's worth. His mettle has been tested.

A fun trip in 2015 to Melbourne to see *Les Misérables* turned into an episode that triggered anxiety and depression. 'Obviously it had been building and I didn't recognise it. →

'I don't want to bag Melbourne, but it was cold and dark, the skies were grey, and our view from the hotel was buildings. No grass, no trees. I figured I'd be okay once I was home, but it never got better.'

That was a very bad year. Ben took over. There was counselling, and medication, and sun-bright walks on the beach. And, always, there were her chickens. Originally obtained to provide fresh eggs, they've morphed into a vital part of her mental health and happiness. 'I came out of that experience with a strong sense of who I am as a person. That's when I connected to my love of chickens. It's something for me that's separate from work and my family and being a mum. It's my thing, my little hobby. They give me purpose. They bring me peace. Some people may think that it's strange, having so much love for your chickens, but I do. I really see them as my pets. It gives me joy looking after them.'

Life is good now. Simple and very good. Cassie notes that her parents have come around. 'They're happy too. I work in a supermarket and my dad couldn't be any prouder. He comes in and sees me serving customers, and tells me, "Oh, you look beautiful behind the counter there," and, "If I was your boss I'd be giving you a raise!" I guess when you're younger you just do things to try and please them, don't you. And now I'm older I have that confidence to do what I want to do.' ●

When Experience Matters.
G.JAMES

SECRET CHICKEN BUSINESS

RICK WASHINGTON

Purveyor of aluminium window frames and doors, private potterer, resting adventurer

Lives: Brisbane, Queensland

Pictured with Little Red Hen (left), a Rhode Island Red, and Blackie (right), an Australorp

'The chickens come first. And then second. And then maybe me, and then maybe one of his children. Um, no, sorry. Before me come the dogs. So there's the chooks, then there's the dogs, then there's me – I think I'm third. And then the hierarchy goes down from there.'
Rick's wife, Karen

NUMBER OF CHICKENS:
Fourteen (hens)

BREEDS:
Araucana, Australorp, Rhode Island Red, Silver Spangled Hamburg ('She's the only bantam. Turns out she's quite bossy.'), Wyandotte and a mystery hatch.

NAMES:
Little Red Hen and Blackie

OTHER PETS:
Button the cavoodle and Artie the attack dachshund

When Experience Matters.

'When he's talking about his chickens he gets a funny glazed look on his face. A glazed-eye look. He thinks they're lovely. Did he tell you how many chickens he has? He fibs. He straight out lies.'

For a minute, maybe two, Artie the dachshund has Blackie in a death stare. Locked on like a tractor beam. It's intense. The dog wants that chicken. Fortuitously for Blackie, one of Rick Washington's two favourite chooks, she's securely tucked under Rick's arm and, while Artie's but centimetres away as the dachschund flies, he is behind a serious, Rick-built fence. Rick is devoted to his chickens, and Artie has been flagged as having, shall we say, impulse-control issues, so the spacious, subtropical, multi-level backyard has been firmly divided between the two classes of beloved pets, giving equal, yet separate, dominion. A fifty-fifty split. 'People don't realise that dachshunds are bred to hunt badgers,' Rick explains, with a flicker of pride at Artie's murderous focus. Rick, seventy-six, and barely winding down after a big, rollicking life in Papua New Guinea, seems to dig the unpredictability that Artie brings to the mix. 'He's a good boy.'

Artie, to be honest, is not so much 'good' as expertly quarantined from temptation. Later, as Rick carries Blackie into the house, the lithe little hunter makes a quick leap and nip at her foot, and draws blood. Rick searches out a tissue, unfazed. Blackie, after all, is fine, and Rick's accustomed to trouble-shooting a menagerie of unpredictable pets. The Washington family lived in Port Moresby before moving back to Brisbane in 1996. According to Rick's daughter, acclaimed singer/songwriter Megan, his choice of family pet was first and foremost chickens – always a flock of ten or more – then whatever else could be drawn from the available pool of Papua New Guinean wildlife. 'There weren't many rules. He had all kinds of things. We had sugar gliders that lived in a coconut. Mice. At least three dogs at a time. A blue-ringed octopus. A baby croc, for a minute. Chickens are about the tamest thing he's had.'

Rick, who also admits to turtles, parrots and lorikeets, acknowledges that the blue-ringed octopus was an overreach. 'It would get itself out of the glass tank of a night-time, and be on the floor the next morning. With two young children we decided we'd better take it back to the beach.' →

Rick has two constants in his life: chooks and a bring-it-on attitude to opportunity and experience. Apprenticed as a baker and pastry chef in Cairns, at age twenty Rick wrote to a bakery in Port Moresby telling them, essentially, what an asset he'd be. Three weeks later he was in PNG, where he stayed for thirty-three years. His second job was as a sweets and pastry chef on an oil rig, then his passion for sound recording turned into a two-decade-long career working as a stringer for Fox and Movietone, and making anthropological and training films for the Department of Information. His wife, Karen, another constant, was equally intrepid when he met her: 'One of the first females that went to PNG, to manage a freight company.' Rick's still working full-time, selling windows and doors. 'Why would I retire? I'd just have to find a job.'

The chickens are Rick's hobby. His mental health counterweight. 'They're peaceful, to me. I go down there sometimes and just watch them. See who's boss. I'll have a chat – "Stop bullying your sister," that kind of thing. It's soothing to my mind when I'm down there, amongst them.' Time vanishes. Rick, engrossed, disappears into his hens; he builds for them, extending their run, double-meshing their enclosure, designing and building accessible laying boxes, or installing a water tank or maintaining the Berlin dog wall. He hoses down the poo, feeds them treats, checks they've not flown into the neighbour's property. He plans, he executes. 'He's gone for hours, just fiddling about,' says Karen. 'I want to get a walkie-talkie, you know, and put it in his pocket, so that when I call him he answers.'

After forty years together, Karen and Rick also have clearly delineated dominions. 'Karen has her sewing, whereas I have my shed down there in the chook yard,' Rick says. Karen rules the roost inside, with exceptional taste, a colonial bowerbird approach to decorating, and a lifetime's collection of treasures. The house is unexceptional from the outside, but inside has a lush, equatorial feel, all decks and levels, plants and carvings. A statue of an ibis, a narrow-waisted kundu drum, a lacquer cabinet, so many bowls. Outside is all Rick's, where he vanishes to tinker with his flock and shed, the latter functionally crammed with tools and benches: a paradise for a certain type of man who likes to saw and build and create. Sometimes he gets carried away and there's conflict between the two fiefdoms. He may, for example, wander off with a 1960s hardwood curtain rod, which came with the house and was only down for a curtain fitting. It's possible he chopped it up without asking and used it for a chook roost. That was a challenging day.

Karen views Rick and his chickens with bewildered affection. 'When he's talking about his chickens he gets a funny glazed look on his face. A glazed-eye look. He thinks they're lovely. Did he tell you how many chickens he has? He fibs. He straight out lies. He'll say fourteen, then I'll go down there to get eggs, and count and say, "There's nineteen there," and he'll say, "Oh, yes, I picked up a couple at Patterson's. I didn't count them." The numbers are fluid.' She pauses. 'It's like when you're a sewer, you never admit to how much fabric you have because people will think you've got a mental illness.'

Rick's always come as a package with chickens, and he always will. Life's unimaginable without them. Occasionally he brings them into the house as a conversation starter. 'I don't do it much. It's more if we're having a dinner party, and we're talking about the chickens, and I'll go and get one. When they're sleeping you can just pick one off the roost.' Karen notes chickens are just part of him. 'Ever since we've been together, which is several thousand decades, he's had them. So maybe the reason he's not very stressed is that he does have the chickens.'

There is always the sense, though, that despite the boundaries, the fences being tended, there's always the chance that they'll give. 'I feel as equal about the chooks as I do about the dogs. The dogs sleep on our bed and if the chooks could, they could. But unfortunately fourteen of a night might be a bit much.' He might want to run that past Karen. ●

G.JAMES

THE PECKING ORDER

A NEVER-ENDING SOAP OPERA

A lot of chicken owners, particularly the ones who care, find watching the pecking order in action distressing. It can be. It's business as usual for chickens but looks an awful lot like bullying from where we stand, featuring lashes of exclusion backed up by forceful pecking. It's generally an acute rather than a chronic issue, something that spikes when you introduce new gals to your flock and they have to negotiate their place. Still, it can be ugly, violent and disconcerting. We humans are bewildered that our beloved and hilarious Betty or Nugget or Germaine can be such a bitch on wheels. We certainly can't work out why poor Cleopatra or Laverne or Countess Chookula is so unpopular. We see it as an HR issue that needs solving: how do you make it stop?

You can't, is the short answer. Chickens give zero clucks what we think about their class system and how they enforce it. If you'd like to experience futility, and who wouldn't, try helping out the chicken on the bottom of the pecking order. There are a whole lot of hours you won't get back.

This doesn't stop us having a crack, of course. I, for example, am made of 100 per cent bleeding heart. It kills me to see any of them miserable, and I've been known to get very cranky with chickens who do not let other chickens eat. The mean girls just look at me as though I'm mad, and get right back to doling out the chook equivalent of wedgies and picking Ms Least Likely last for the netball team. They can be a breathtaking pack of molls, sometimes thugs, and for anyone who experienced bullying at school (spoiler alert: me) it's pretty triggering.

The good news is that it mostly sorts itself out once the hierarchy is re-established, but you will never have a conflict-free flock, or no unpopular chickens. That is not how chickens roll. For chooks, the pecking order is a matter of life and death.

The pecking order is how a flock orders itself. There's always a top chook, a bottom chook and, in between, a cascading hierarchy enforced by pecking down. The middle section, in particular, is fluid, and there are inevitable skirmishes when new chickens are added, a top chook dies, young chooks grow up, others get old, a fox intervenes, a chicken gets sick or new alliances form. The pecking order is enforced, and necessary. Chooks need to know who is in charge. Who eats first, then second and so on, because otherwise every meal would be a fight, and they'd exhaust themselves competing with each other. Chickens are tasty, tasty prey animals, and if they don't work together they're toast. Counterintuitive though it seems, the pecking order is chooks working in concert to protect each other, and the flock. Even the ostracised bottom chook is doing her job.

It took me a while, when I first got chickens, to understand that they would never be communists. I went in naively, believing that if there were enough resources, an abundance of food and space and treats, they would stop 'being mean'. They would not. Chickens will never share equally. They will always find another chicken's treat more desirable than the one in front of their own beak. And they will always warn off the chickens who are lower down the pecking order. →

If there were eighteen piles of sweet corn and mealworms raining from the sky, the top chook would still be duty-bound to body-block the other chickens from all of it until she's full. Otherwise she is no longer top chook, no longer doing her job, and chaos will ensue. The pecking order is their survival mechanism.

Depressingly, it's this social sophistication, this drive to connect and negotiate, that has enabled humans to successfully exploit chickens through intensive farming. And, also, why it's so cruel. Chooks, ideally, live in flocks of twenty to thirty. They all know each other and who they are in relation to each other. When they're in claustrophobic quarters with, say, 10,000 other chickens, each day is an endless battle for hierarchy with an infinite supply of manically stressed strangers up in their business. It's literally a chicken's worst nightmare.

...

What about roosters? I hear you ask.

They provide a vital service to their flock, taking on a lot of the leadership roles: finding food, keeping an eye out for predators and threats, flirting, keeping the gals safe (a good rooster will take on a fox to protect the hens). But roosters have their own hierarchy, and don't have a position *within* the flock. They do, however, keep the hens in line, and the lid on mean-girl behaviour.

These days, of course, roosters are like live-music venues – no longer welcome in the inner city or suburbs – so backyard and commercial flocks don't have them. Their traditional service is considered a disturbance of the peace and something to complain about, rather than a stamp of authenticity and vitality.

Ah well. A flock of backyard hens mostly gets by just fine without a roo. Except they get bitchier.

This, our third flock, has been mostly chill. The girls move up and down the ladder of power, dominance and popularity without issue. Joan usually lurks somewhere in the middle: small, a little eccentric, but with a lot of fizz. I've always loved her for her overbeak, which gives her an aristocratic air.

Recently, though, her standing in the flock has plummeted like a watermelon dropped from a third-floor window. Last year's chicks are all grown up, so naturally they're challenging the establishment. 'Why should I have to wait my turn behind Too Old Too Slow over there?' I see young Dora thinking, as she pulses with hormones and self-righteousness. Dora is a self-important chicken with a lot of opinions who is going places. I note that Dora hasn't gone after Phyllis, the Silkie who is currently top chook. She's not stupid. Joan, the smallest of the chickens, who is prone to respiratory infections, is low-hanging fruit. She's also friendless because her sister Violet, the moll, has aligned herself with the twin Silkies. Joan is increasingly anxious as Dora's reign of terror escalates and the others, including Violet, join in. Chickens learn through imitation, and savage pecks to the back of the head are trending.

Joan is living as a hyper-vigilant fugitive. When dawn breaks she's up on the roof of the coop to avoid the bullying; at night she's the last in, dillydallying to avoid the inevitable 'covfefe' of finding a perch, pretending that the garden at dusk is oh so fascinating and it's not a half-hour past her bedtime. She hides, solo, under a bush to avoid the others. She tries to perch at night in the garage. We'd let her, but we can't guarantee she's safe #foxes. This has gone beyond regular pecking-order stuff; it's out and out persecution. The tribe has spoken, and Joan, for now at least, is out of the flock.

Chickens know things we don't. They can see in ultraviolet. Who knows how Joan appears to them right now? I suspect not great. To human eyes she's a goofy underdog chicken with a sweet personality and a passion for sweet corn. But the weak get weaker in the world of chooks, and she's always been a bit NQR. She lays soft eggs, gets frequent 'colds' and moults often and almost in cartoonish fashion – you know, *poof* and her feathers are on the ground. I think, from their perspective, she's become a liability. What to do, what to do.

Joan's full of fear, either from the bullying or from the knowledge she's being cast out. She's no longer fulfilling her role in the flock; indeed, now that they've ostracised her, she no longer has one. A chicken rejected by her flock is not going to make old bones. Neither will a chicken who sees no future for herself.

I've already had a chicken will herself to death. Bunty was in our second flock, a timid Australorp bantam, and she was monstered by twin Araucana crosses who had the personality and general vibe of Death Eaters. They were relentless, and I watched her give up, stop eating and die. We weren't as sorry as we might have been when the fox came for the rest.

We are trying, lord knows, to help. We're making sure Joan eats. We're letting her out to free range and leaving the others locked in the run to raise her status. We're shutting Dora and Violet in the broody cage, to break the cycle and lower their aggression. We've discovered that yelling, 'Stop being mean' doesn't achieve anything. (To be fair, we already knew that.) We've trimmed her beak, so she can use it to retaliate. We discuss strategies: give Dora away? Give Joan away? (Who would want that mangy, anxious, tiny darling?) Buy a supply of chicken nappies and make Joan a house chicken? We are attempting to social-engineer our flock's pecking order, and it's a full-time gig. We are seeing incremental improvement, which keeps us in the game. The tricky part is that we simply don't know all the rules. Chickens are not beholden to our laws or preferences. They do not root for the underdog. The flock is paramount, and there is no room for sentiment when you're prey.

...

I'm still rooting for Joan. She hasn't given up. She has pluck. But this is why chickens are addictive: the drama. There is always something going on. It's the pecking order that makes them magnetic, a soap opera which runs and runs. There are heroes and villains, alliances and violence, friendships, clowns and crazed loners. There are characters and power plays, arguments and feuds, births and deaths. Whenever you tune in, there they are: always familiar, always high stakes, never predictable. Because of their intricate social hierarchy, we see ourselves reflected back. Despite best intentions it's impossible not to anthropomorphise. Chickens engage our emotions in complex ways. A dog loves us unconditionally, and that's great. And a cat ignores us theatrically, which is also great. But chickens . . . chickens put on a *show*. They do the tragedies, the stories of revenge, the comedies, the dramas – and, if you get a rooster, the love stories which range from *Gone with the Wind* to *Mr Bean*.

There is another drama, a subplot if you will, unfolding within our flock. Blondie, another of the young Rosecombs but lower down the pecking order, attacks Priscilla when she gets a chance. Priscilla goes broody a lot (#SilkieLyfe), and a broody chicken is a befuddled chicken. Blondie goes in *hard*. Beak full of Priscilla's pompom, *bam bam bam*. This looks like a straight power play, except that it was Priscilla who tried to kill Blondie when she was a chick. Priscilla's the reason Blondie's neck has a ring of scar tissue that gives her the air of a hipster dabbling in tribal branding.

This is sweet revenge, pure and simple, a story that could have been ripped whole from the pages of Greek mythology. Or, you know, *Survivor*.

Chickens. They are *epic*. ●

LIGHTNING IN A JAR

SUMMER FARRELLY

Creator of Chickens to Love therapy program, artist, community advocate superhero

Lives: Bundaberg, Queensland

Pictured with Ralph, an ISA Brown

'Everyone has the ability to shine. Some of us just shine from a different angle.'

NUMBER OF CHICKENS:
Ten (home flock, hens). Many more in the therapy program, including Ralph.

BREEDS:
Belgian d'Uccle, Frizzle Polish cross, Modern Game, Pekin (black, millefleur and white) and Polish

NAMES:
China, Crow, Zoe Zoe, Destiny, Pink, Tango, Fuzzle, Domino, Nang Hei and Splosh

OTHER PETS:
Two cats, Katie and Winter, and four lizards, Norbert, Mr Floppy, Spike and Cubular

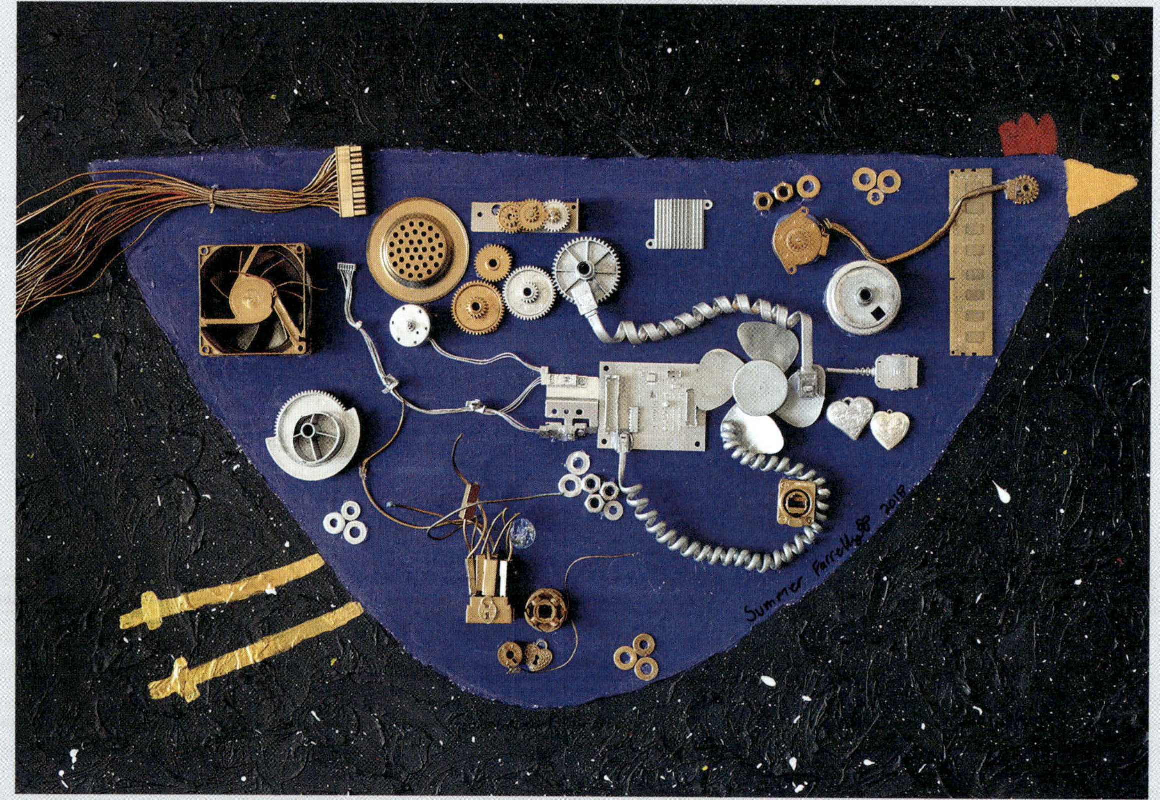
Summer Farrelly 2018

Twelve-year-old Summer Farrelly's 'Cyborg Chicken' multimedia artworks are striking. Probably why they sell so well. They're good. Oils, I'm guessing – thick colours and gobs of texture for the feathers, and eyes made from old camera lenses and phone innards. All the better to examine you with. *Whirr click*. Her chickens are augmented with dismembered machinery: studded with cogs, gears and wires, a few tea strainers, the spattered remains of dead tech sprayed gold or left as is after being cannibalised. It's a total steam/cyberpunk overlay on two-dimensional, front-facing, bright, friendly chickens. Models drawn from the artist's own chooks. The fusion works; there's synergy here. And what better way to express, actually, what it is to have ASD (autism spectrum disorder), and how it is to navigate the world through the eyes of a chicken?

Summer, bright as a star, obsessed with chickens, is explosively creative. She leads from the heart and gathers people in her wake (Mum Cynthia – her greatest supporter – included). Her projects are legion. She's drawn together a 3D printing team to design and make prosthetics for chickens that have lost their feet or legs, and she's creating a poultry leg and beak database. She posts regular 'chicken enrichment' videos, inventing and demonstrating devices such as a 'bread chandelier' to help chickens enjoy life more. She's distributed fifty sensory packs around community centres in Bundaberg (funded by selling her art); each contains noise-cancelling headphones, sunglasses and distraction tools to help people with autism manage their sensory challenges in crowds and at events. For her eleventh birthday she asked for a street library, so she could share books. She's a legend in her community; everything is about giving.

'I don't want things for myself,' she explains with enthusiasm, sitting at the kitchen table. She brushes a crumb of buttered breakfast toast from a crease in her 'I have O.C.D. – Obsessive Chicken Disorder' T-shirt. 'Most kids are normally like, "I want this, I want that," whereas I normally say, "I want this so I can give it to that." It's the happy feeling I get knowing that it will satisfy someone, make someone happy. We have a big mandarin tree, and I always bag them and give them away. I barely keep any, because I know I can get some any time, really, just by buying them. But other people can't.'

In May 2019 Summer was Triple M Bundaberg's 'Little Legend' of the week, a radio promotion that gave $250 Harvey Norman vouchers to kids who achieved something outstanding. Brekky announcers Lu and Matt asked what she'd do with the prize, and Summer said she'd already decided: 'My brother's starting a photography club at his primary school and I'm donating it so they can buy some equipment.' Clearly, no one had ever done this. Lu said, 'Oh my god,' under her breath, and then there was a radio silence you could drive a truck through. 'Who,' you could hear them thinking, '*is* this kid?'

Short answer, Summer Farrelly's the gal who, at ten, co-wrote a world-first autism therapy program called Chickens to Love. The program uses chickens as therapy animals to help those on the autism spectrum develop social skills and self-esteem, based on what she learned from her own chickens. 'It's all about spreading the chicken love, and creating a place of acceptance and a place to belong.' The Chickens to Love compound, abutting lush green fields of Queensland cane, sits on land provided by Bundaberg's Community Lifestyle Support centre. There's a large coop with disability access, and a flock of ISA Browns hand-socialised by Summer. Wanting to make the point that no-one should be judged on their looks, she chose ISAs because they're all similar: 'I didn't want any chicken to stand out and attract more attention.' There's also a shed stocked with chicken enrichment activities and 'autism is my superpower' capes, all used for community programs. And a Zen garden, landscaped by Cynthia. It's beautiful. A peaceful place with a Buddha water fountain where the therapy chooks can forage. Chickens to Love has been running since 2017, and there's interest in rolling it out nationally.

The program came from Summer's own struggle to learn social cues, and how it impacted her at school. Being on the autism spectrum brings with it a suite of traits (or as Summer now calls them, 'super powers!'). →

Summer's diagnosis used to be known as Asperger's syndrome, and the characteristics that commonly accompany it include having one fixed special interest to the exclusion of others (think Greta Thunberg and climate change – or Summer Farrelly and chickens), and being unable to decode body language, tone or facial expressions. This translates to taking things literally, and having to actively learn how to read people. When is someone pleased to see you? When do they wish you'd rack off? It's a mystery. The school terrain is treacherous for anyone; it's deadset impossible to navigate the roaming packs of cool kids and cliques if you don't comprehend sarcasm or jokes. Playground dynamics were utterly lost on Summer, and her obliviousness did not make her popular. It was a bad time.

She spent countless hours outside with her only friends: her lovely chooks. She did a deep-dive into the flock and found her emotional equilibrium. She learned how to trust again, how to love and be loved back. 'Chickens don't discriminate,' she explains. 'They love everyone. They made me feel happy, loved, needed and special.' She gradually started to notice and pick apart the flock dynamics. The social hierarchy. The body language. The way the chickens arranged themselves into smaller flocks (friendship groups!), the top chooks (mean girls!), and the chooks that were bullied (Summer!). Wait a minute. She mapped the pecking order onto the social dynamics of the playground and it clicked into focus. 'If I pretend the children at school are chickens, I can understand social situations.' It was a revelation.

'If I put down a piece of watermelon, there were dominant chickens who would run up and push everybody else away. They reminded me of the playground. They were the kids who would claim a spot, and no one else is allowed in it. And there are the chickens who are too frightened to even try – I make sure they get something off to the side – and maybe in the middle you'll get a nice chicken who'll let in anybody. I look at kids at school now and go, "Oh, okay, I'm not gonna try and play with them because I know my answer. All I'd do is just make more enemies."'

Chickens had always been Summer's thing. It makes sense. They're soothing company; when you're on the spectrum, and sensory overload is real, the quiet company of a flock of chickens is a haven. But Chickens to Love kicked everything up to a whole other level, gathering the entire family under its wing, providing a much-needed common interest. Summer's mum has an ASD diagnosis, as do her brothers, Oliver and Findlay. They were battling for connection, and there was a lot of chaos.

'I don't think we would have connected at all, without the chickens. We'd still be stressed out twenty-four seven, with nothing in common,' Summer says, noting that her brothers took on chook-related tasks. 'Oliver would grow the barley sprouts, Findlay would breed mealworms. It gave us something to talk about. Oliver takes the photographs for my Cyborg chicken cards. And Findlay doesn't like to admit it, but you can see him, he'll go out in the garden and be like, "Hello, chickens." He pats them.'

Passion for chickens and inclusivity have transformed Summer into an advocate for autism and a powerful community activist. She's lightning in a jar. She was Bundaberg's Young Citizen of the Year in 2018, one of the Queen's baton carriers for the 2018 Commonwealth Games, the winner of Autism Queensland's Creative Futures Recognition Award in 2018, the 2017 winner of the ABC's Girls Change The World contest, and she's been featured on *Australian Story* and the web series *Girls On Fire*. Not, to be clear, that she would mention *any* of this at school. Heck, no. 'If the teacher mentions me, says I've won an award or something, everyone looks at me and I'm like, "Aw. Flibble-flabble. This isn't what I needed,"' says Summer. 'They're just going to make a way to try and insult me.' Knowing you're bottom of the pecking order doesn't necessarily mean you can change it.

The chickens are always there for comfort, whether it's the Chickens to Love flock or Summer's home flock of soft fluffy bantams. Like Crow, the black Pekin, who sits on her head like a Russian dowager's hat. 'They're something that I can go home to and know that they're not gonna be rude to me. I can just sit there and they'll come to me and I can chill with them, play with them. They're not gonna yell at me or get angry.'

Even as her vocation as a community, chicken and autism activist soared, Summer was still being bullied at school. The difference now is that she understands why, and doesn't particularly care. She knows she doesn't fit in. 'I'm not all, "make-up this, make-up that". I can't even name a piece of make-up. There's a group I called the "girly girls", and every girl was in that group except me. And they'd all be like, "Oh, I got my nails done! And I like texting!" So, yeah, most of the people I'm making friends with are adults.'

In 2018, though, the bullying escalated again and Cynthia decided to home-school. Summer's parting shot was so *her*. All class and heart. She's donated a 'Community Spirit' award to her ex-school, to be presented each term as recognition of a student's achievement outside sports and academics. 'I want to build self-esteem within the school community,' she said on the Chickens to Love Facebook page. 'There are so many students who are not academic, but have amazing abilities that are community minded, creative, innovative, caring and artistic.' →

Community activism is one way Summer's found to negotiate and thrive in the neurotypical world. All courtesy of her chooks. She's also developed a cracking sense of humour. 'One thing I do,' she says, mischievously, leading up to the reveal, 'is keep my hands in my pockets. Then, when people are being mean to me, I can flip them the bird and they have no idea.' ●

I HAVE
CHICKEN
Snake skin

BACK TO COUNTRY

JOSEPHINE WOODS AND ROSIE PATTERSON

Patients at Purple House

Pictured with Dame Eggna Doris Hermione Napaltjari, a White Sussex

CHERIE HUGHES

Logistics manager

Live: Alice Springs, Northern Territory

'Patients give the chickens skin names. They want to have the chickens named after their skins.' Cherie

NUMBER OF CHICKENS:
Three (hens)

BREEDS:
ISA Brown and White Sussex

NAMES:
Dame Eggna Doris Hermione Napaltjari, Astro Alex Henny Penny Nangala and Napanangka

OTHER PETS:
A dog, Pretzel

'We have all sorts of visitors and volunteers . . . and I think they walk outside and see the chickens and it gives them some kind of breathing-out thing. Like, "Oh wow, this is *nice*."'

The headquarters of Purple House sits, unobtrusive, on a suburban street on the outer edge of Alice Springs, right in the dead red centre of 900,000 square kilometres of desert. Purple House is an Indigenous-owned and -run dialysis support provider, servicing the largest renal failure catchment area in the world. On Radio National the Purple House CEO, Sarah Brown, described Alice Springs, without any hint of exaggeration, as 'The dialysis epicentre of the universe.' A significant chunk of Alice's population is reliant on dialysis.

Desert mob – Indigenous people from remote communities – experience renal failure at up to thirty times the national rate. And when they do, they must move off their country, travelling hundreds or thousands of kilometres, and come to Alice for treatment. Dialysis is exhausting and restrictive: three days a week, most likely forever. They can't go back home, not with treatment every other day. They miss their family, their land, their language, and they're usually living in a hostel: dislocated, isolated and depressed. Not great for health outcomes. The impact ripples out. Because sick elders leave permanently, their knowledge and leadership are lost to their communities. Families are separated. The culture suffers. From the community's perspective, someone becomes ill, they go to Alice, they die. It's a catastrophic lose–lose. Purple House was imagined as a response to this, a provider of dialysis that serves remote communities and keeps people connected. It's somewhere for Alice-based Indigenous dialysis patients to call home.

And, oh, it is a lovely place. Warm, welcoming and arty, with colourful walls, a fire pit in the garden and a large Aboriginal flag hung proudly on a fence. A sizeable painting leans against a purple wall, and an effigy of much-loved local Labor pollie Warren Snowdon is propped jauntily outside, maybe losing a bit of stuffing. It says 'hippy cafe' as much as it says 'clinic'. Purple House has two dialysis machines and provides medical support, but this is given alongside shopping trips, music sessions, picnics, bingo and the chance to do your washing. Staff buzz around, offering shawls and socks. There's an upbeat casual vibe. The place feels lived in. When you've got end-stage renal disease (ESRD) and you're forced away from your community, Purple House is everything. →

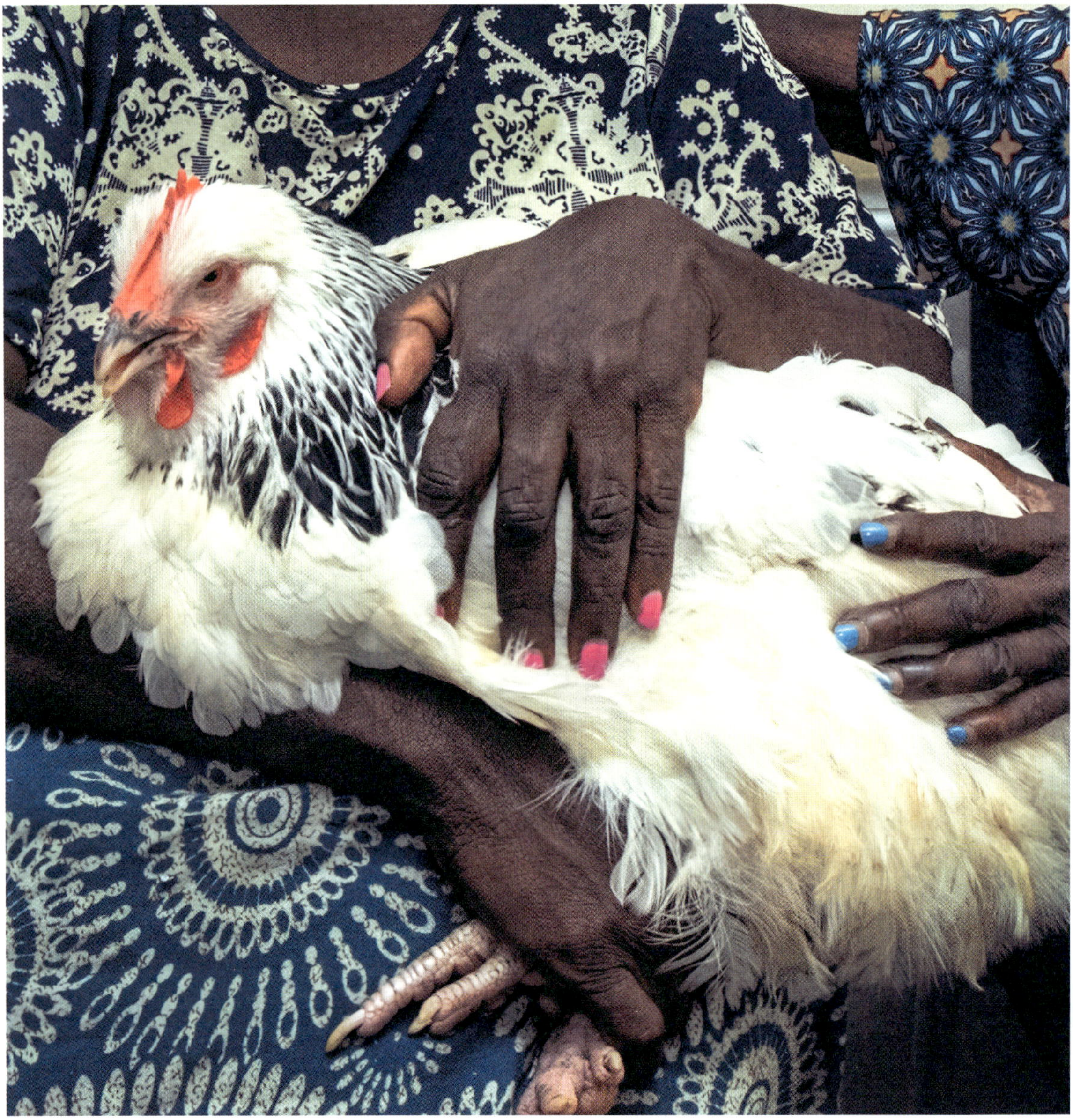

The homeliness is upped a couple of notches by three friendly chooks in their large artisanal coop, located in the garden, a sensible distance from the fire pit. They have roast kangaroo-tail feasts here, but the patients often joke that the chooks would 'make a good meal, eh'. Dame Eggna Doris Hermione Napaltjari is a plump White Sussex (and winner in the category of Sussex [Female] Class P14, Heavy Breed Large Fowl at the 2017 Alice Springs Show), so it says something about the chooks' status as pets that they're still here. 'They're the best fed chooks in Alice,' says Cherie Hughes, logistics manager, referring to the kitchen scraps and food brought by visitors.

Purple House HQ contains a thrumming office, which coordinates the Purple Truck (their mobile dialysis vehicle), and the remote clinics in communities from Yirrkala to Warburton, but above all it's a social hub. Only two people have dialysis here each day, but most days of the week there are thirty to thirty-five patients visiting, just hanging out. It's somewhere to come and have a cup of tea and a good feed,

meet your friends, have a shower, get your hair done. There's an outdoor salon down the side of the house – with mirrors, a free-standing hairdresser's basin and a framed salon-style drawing of a woman gazing out from beneath her long, side-parted hair. Rosie Patterson, an Anmatyerre woman from Yuelamu, and Josephine Woods, a Luritja woman from Papunya, come to Purple House most days. Today they've had their nails painted. Rosie's are a deep blue, Josie's a bright pink. The colours really pop against Dame Eggna's white feathers.

'The chickens is good when the kids come here,' says Josie, sixty-two, who has some family in Alice, and a keen interest in the eggs laid by Dame Eggna, Astro and Napanangka. 'Love eggs. The chicken's good for eggs.' Sometimes she stays in a hostel, sometimes with family. She's been on dialysis for ESRD since 2011. Eight years of not living in Papunya. 'When the kids come here, out they go. See the chickens, play around with them.' She laughs, holding on tight, she and Rosie stroking the placid fat chook together. 'Chickens are lovely. Nice to hold. →

Nice to chase around. We used to have a pig farm, used to run down the creek. There was some chicken. We'd catch them and cook 'em up and eat it. That's what we used to do.' She strokes Dame Eggna again. 'Long time ago now.'

The Purple House chooks don't get chased, bless 'em. Particularly not since they moved into their coop. It's a corrugated iron-panelled arch of steel wire ribs built by the husband of one of the dialysis nurses, open at one side, a painting tucked down the back among the straw, and a mirror ball hung from the ceiling. 'We had them very free range,' says Cherie. 'But there are cleanliness problems with "very free range" in the clinic. They used to have free run, and they'd invite themselves inside of an afternoon. So now we've got their little barn house. But them being integrated is part of the wellbeing aspect of, you know, healthy life, healthy living. Animals create a softness between people, and the chickens are part of that.'

The chickens are part of the Purple House team. Part of what makes it a community hub. A small part of what makes it work. Rosie's in her seventies and has been on dialysis for ESRD since 2009. Ten years. The health outcomes for desert mob on dialysis used to be the worst in Australia. They were dying two and a half years after leaving for treatment in Alice Springs. The homesickness. The Pintupi people from the Western Desert wanted a new model of care, based around family, country and compassion, and decided to do something. In 2000 Papunya Tula artists from Walungurru and Kiwirrkurra developed four culturally important collaborative paintings, which were auctioned with some other works at the Art Gallery of New South Wales and raised more than a million dollars. Enough to found the Western Desert Nganampa Walytja Palyantjaku Tjutaku Aboriginal Corporation, now known as Purple House. They have sixteen clinics in remote areas, with more planned, which means patients can be rotated back home to visit, to be with their language group and take part in their cultural life: attend ceremonies, footy carnivals, funerals. There are measurable mental and physical health outcomes. Purple House patients now have the best dialysis outcomes in Australia.

Fundraising is ongoing and energetic. They've got their own Medicare item number now, which rebates $590 per dialysis treatment in a remote location, which helps, but demand for Purple House's services is exponential. They currently service about 150 patients, but there are so many more. They're still auctioning art and, as part of the fundraising for the Purple Truck, they held a competition to name the chickens, raising about $2000. It turns out this is why the gals have such grandiloquent names. 'We thought it would be nice if a lot of people won. Share the love around. So, yeah, they ended up with a few.'

The chickens have been doing their bit at Purple House for a long time. Before Dame Eggna Napaltjari, Astro Nangala and Napanangka, there were four Silkies, also with skin names. The chooks' job is to increase wellbeing and bring a sense of community. Do their usual value add. Cherie says they're good for staff wellbeing. 'A lot of the staff get attached. They're out there feeding the chickens, making sure they're all right. We have all sorts of visitors and volunteers – all sorts come and visit – and I think they walk outside and see the chickens and it gives them some kind of breathing-out thing. Like, "Oh wow, this is *nice*."'

With the patients, the chickens give visiting kids and grandkids a focus, and add extra life and dimension to what's already an energetic space. Animals are always part of a community's bustle. And there's a definite component of reminiscence, of chickens caught and eaten past. 'Chicken's a favourite meal, you know,' says Josie. Sitting at Purple House with friends, having a joke about how Dame Eggna or Napanangka would be tasty, is a memory track straight back to home, back to family. I ask Rosie what she likes about chickens. 'I like 'em round the bush. My son, he'd get the chicken. We'd go the chop. My son would do the chopping. We eat the chicken.' She glances at Dame Eggna. 'But nah, not eat this one.' ●

SHARE A TALE

THE JUDGE

KELVIN SMITH

Poultry judge, novelty tie fancier, big personality

Lives: Stanhope, Victoria

Pictured with Best in Show, a Plymouth Rock

'To be a judge you've got to have an appreciation for the fowls and what constitutes the difference between breeds. You've got to know your breed. And the standards. Good birds always pick themselves, you only need a judge to sort the rubbish out.'

NUMBER OF CHICKENS:
At the moment? None. Just sold all 250 of them.

BREEDS:
Hamburg (all colours: Golden Pencilled, Golden Spangled, Silver Pencilled, Silver Spangled, blacks and blues), large Old English Game

OTHER PETS:
Indian Runner ducks

Kelvin Smith, until recently of Shepparton, a large, blunt man in his fifties with a big, ready laugh and plenty of opinions, started breeding purebred poultry in the early 1970s when he was eleven years old. Picked it up at primary school in grade six. Started showing when he was fifteen. Lifetime member of the Goulburn Valley Poultry Club. President for ten years. Been a judge for thirty-three years. For every show he's ever judged, he's got up at sparrow's fart, put on his very best clothes, and finished off the outfit with a lairy Foghorn Leghorn tie. It lends a definite cut to his jib.

'Oh, it was given to me by somebody. Can't remember. It was before any of the women, so I've had it a long time. It's just my little bit of a take on life, which is don't take yourself too bloody seriously. Because I don't. Today I'm the judge, tomorrow I might be an exhibitor. Doesn't matter. I've got my bloody good clothes on, my jacket and my chook tie.'

Kelvin came up through the poultry ranks the old way, from junior to senior. It was a desirable hobby back then. Something that a young boy could do without supervision. Something you could do on the cheap. 'We didn't even have colour TV then. People were lucky to have a TV full stop. Everything was outside. It was bike riding, it was fishing, it was animals.'

More people kept fowls then. Everybody had chooks. Kelvin's father was a successful budgie breeder. It's a different lifestyle now, he notes, Netflix and iPads. Still, there have always been distractions from the true path. Even back in the day, not many juniors stayed the distance. 'I was brought up on a farm in the countryside. I didn't have the distractions of the townies. But in the town, as soon as they discovered what they had between their legs wasn't just for weeing through, the chooks took a back seat and eventually disappeared.'

Kelvin considers himself lucky. 'I never had to get rid of them, as such.' When he went to university, his parents looked after his chickens. Later, before he settled down back in Shep, he struck a deal with another breeder to look after his chooks in return for feed for both flocks. 'I had chickens running through my blood, and for many years life was just about the chooks.' His persistence paid off. 'In the late 2000s I was someone to be reckoned with, you know. I'm pretty sure there were plenty of people that shuddered in their feet when they saw my car driving to a showground.'

Part of his success as a judge and exhibitor is that he learned breeding and judging from the greats. 'I did a lot of listening to chaps who wore jackets and tweed caps - master showmen - I think you get more out of talking to someone who knows what they're talking about than from a book. One of the blokes who taught me was a military cop in the Second World War, big imposing bloke, chickens were his passion. Bruce Trewin. He was a master. He'd walk me around his birds and pick out points. "This is what you want, this is what you don't want," blah blah blah. He's dead now. They're all dead. I'm considered an old guy now. Don't ever for once think that I know everything. I don't. But I've been around long enough to know a lot.'

Poultry clubs are transforming, and the scene is not what it was. Kelvin notes that where you'd once get forty or fifty people along to a poultry club meeting, now you get four or five. Distractions. It's also more expensive to show now. When he started, it was twenty cents to lodge a bird at a show; now it's three dollars per fowl. Wheat's doubled in price due to the drought, $500 a tonne, and 250 chickens eat a lot of grain. 'To be a good breeder you've got to be able to breed numbers. You don't just buy a trio of birds, take them home and breed them one year, and think you're a breeder. It takes years and years of specialised selection. And to breed numbers you've got to be able to afford to feed the damn things.' →

DISTRIBUTED BY ACCO AUSTRALIA PTY LTD
All the leading bra
ACCO

Ducks
Turkeys
Hatching - Rearing Poultry
Showing Poultry
Auctions
Geese
Poultry Nutrition - Genetics
Backyard Poultry
Rare Poultry
Welcome

Times do change, and so do people. Kelvin's still judging, but he recently sold his breeding fowls. His first time without chooks in forty-five years. Bloody hell, Kelvin. He has reasons. His new job in farm supplies means he has to work Saturdays. Exhibiting and judging destroys your weekends. Also, he's in his fifties, there was no money to be made on the farm and it was getting to be a bit of hard work. He's got a new wife, no kids. He sold up and they moved into town. 'Just got a little bloody backyard now.' Still, it suits them. 'We can do what we like. I'm not spending every weekend going to shows.' Kelvin goes boogie boarding, down at Ocean Grove. He collects antique bottles, and now he can go to bottle shows. 'I do miss it. But I enjoy my new life a bit more.'

Never say never, though. He didn't sell his big trailer, and it still has a shed on it. 'I can see it across the yard. You never lose the interest. One day I might see something that tickles my fancy. I've enjoyed the whole experience. Maybe one day I'll get that last shed down, and set it up. I didn't sell all my incubators. I didn't sell all my brooders. I didn't sell all of my respectable show pants and I kept one shed. So. What does that tell you?'

In his blood. ●

MS FOX

THE DARK SIDE OF CHOOK OWNERSHIP

You're not truly a chicken owner until you've had a flock taken by a fox. It toughens and transforms you. Foxes are the dark side of chook ownership, and when you lose your beloved gals to this apex predator – and you probably will – you come out the other side a different human. I had no concept of the word 'massacre' until a fox went through my first flock. Betty! Missus! Shirley! There's no unseeing it.

I guess I'd thought they'd take one for dinner and that would be one less chicken – sad but tidy, like an illustration from a storybook of a guilty-looking fox carrying a roast chicken. Not so much. Foxes are the violent spree killers of the animal kingdom, ripping the entire flock to chunks and pieces, apparently for kicks. Technically, they're classified as 'surplus killers', which means they are indiscriminate and slaughter more than they need. Way, way more than they need. They eat and keep killing.

Think of them as the Joker. That gives you more of a sense of what you're up against. It also makes you Batman, which is kind of cool, and your chooks the innocent citizens of Gotham City who need protecting. Batman, obviously, has to be ever-vigilant, because the Joker never, ever stops coming.

The red fox is, like all of the destructive species in Australia, invasive. It was introduced to Australia between the 1840s and the 1870s by a conga line of utter muppets in Victoria, wealthy settlers keen to indulge in the 'noble sport of fox-hunting'. Gah. By the early 1900s good old *Vulpes vulpes* was declared a pest – entrenched and established, no doubt, due to a plentiful supply of that other introduced varmint, rabbits. If you ever stumble across a time machine, please go back to 1874 and persuade the Chirnside family of Werribee Park – whose fox-releasing antics succeeded where others failed – to take up a less destructive hobby, such as stamp-collecting or sand-papering off their own faces. →

THINGS YOU MAY NOT KNOW ABOUT FOXES

1

Like coconut on a lamington, they're all over Australia, apart from the far north and Tassie. If only I had the proverbial dollar for every time I've heard, 'Oh, there are foxes in the city?' Yeah, baby. Foxes *love* the city. They're highly adaptable and opportunistic omnivores, scavengers that flourish alongside humans and are the font of all garbage. In the suburbs, there can be up to sixteen of the sensationally cheek-boned russet predators per square kilometre. Foxes live along creek beds and train tracks, below uninhabited houses (thanks, negative gearing), houses with only one person, vacant lots, parks, reserves, anywhere there's grass or soil or piles of undisturbed lumber or abandoned building rubble or other crap.

A friend has seen them at dusk as he jogs through our large, local, very inner-city cemetery. The mind boggles as to exactly where the dens are.

2

Of an evening, foxes patrol places of interest. Once they've sussed out you have chickens – and, boy, can foxes sniff warm-blooded, delicious, snoozing chicken on the wind – you're on their beat. They visit once or twice a night, every night, just checking in. Patiently waiting for the one time you get home late and leave the coop door unlocked. Or for when your dog's away for the weekend. Or for when shifting soil or weather or weathering of materials allows them to flow like mercury through a gap, under or over the coop fence.

Or, in my case, for the day I was trying to be clever and thwart the pigeons who were flying into the run and eating the chicken feed. I closed the door while the girls were out free-ranging and forgot I'd done it. Human error. In retrospect I realise we were a bit cavalier. We'd not had any problems with foxes; we figured we'd got it all sorted. The gals went to sleep outside, and that was the first time I lost a flock. Oh wow, did we hate ourselves for a while after that.

The second time was about eight months later. We'd done a hasty restock around our one surviving chicken, Laverne. My mum had died, we were distracted, I thought Greg had locked them in, he thought I'd done it – you can see where this is going. Bam. We took a break for a year before we restocked again. We needed to let the grass grow in the run, let the bad vibes dissipate. A jungle of plants sprouted, ferns and tomatoes, wheat and pumpkin, all from buried and leftover chicken seed.

It was the right move, and we love our new gals.

Last month we were woken by a weird, hollow howling at 4 am, and there was Ms or Mr Fox being held at bay in the driveway by our cat, Harry, who was most definitely not meant to be outside. So there you go, we're back on the list. Harry scooted away when we came outside, but the fox didn't run. Just jinked past us into the garden to check out the coop. The cheek! Bold as a hens' night at a male strip joint.

3

The red fox has been fingered as a major player in the extinctions and near-extinctions of many medium-sized native species. Your bandicoots and bilbies, rat-kangaroos and quolls, your hare-wallabies. See 'surplus killing' on the previous page. In Australia the fox is a 'keystone species' which greatly reduces biodiversity, and is considered a threat to fourteen species of birds, forty-eight mammals, twelve reptiles and two amphibians. They've been known to chew the teats off cows, and in the Mallee they eat endangered Malleefowl eggs and young. As the cherry on top, if (when) rabies ever makes it to Australia, foxes will be a dominant vector of transmission. Ugh. Foxes are the *worst*. →

Look. It's not inevitable that a fox will get your chickens. But it's likely. It might not happen for years, but I don't know anyone who hasn't lost their chooks, at some juncture, to a fox. We've all got war stories. There was a gap in the fence. The neighbour was looking after them and forgot to lock them in. The dog died and 'nek minnit' the fox turns up. They dug a tunnel from next door. They chewed their way through the wooden coop frame. They hijacked a plane, flew over the run, and parachuted in.

There are chickens in this book who've already been killed by a fox. It's bleak, but important that you're forewarned, so you can recover when it happens. The fox is *designed* to get your chickens - that's pretty much its prime directive - so be kind to yourself when the *merde* hits the fan.

It is awful. And you grieve. And you make a note to self about how to not make the same mistake again (forget the pigeons!). You're stronger and wiser. And then, when it feels right, when you miss the gentle susurration of chickens *bok-bok-bokk*ing in the garden, you pull up your big farmer undies and go to the poultry auction and buy more chickens. Or a friend will give you some of theirs. Or you'll rescue some battery hens. The next flock will be different from the one you lost. You never get those chickens back. But there will be new cuties and stand-outs with different personalities; you'll buy a breed you've always fancied but didn't have room for; there will be new problems to solve, and you'll build on the lessons you learned in the past. And there you are: a true chicken owner! Moving forward.

And you are. Because having chickens is important, and life-affirming, and one of the best things you can do for the planet. And if you don't restock, well, the fox wins. And we can't be having that. ●

PAGAN POP PRINCESS

ADELE SCOTT

Interior designer, feminerd, burlesque artist, pagan

Lives: Albion, Victoria

Pictured with Spike, a Frizzle

'Sometimes the rebellion is being joyful in a very sad world.'

NUMBER OF CHICKENS: Currently? Zero. Chickens that were? Nine. Chickens that will be again? So many.

BREEDS: Araucana cross, Belgian d'Uccle, Black French Marans, Frizzle, Golden Laced Wyandotte, Old English Game (hens)

NAMES: Spike, Lemon Meringue Pie aka Meringue, Sticky Date Pudding aka Pudding, Drucilla and Brucilla aka 'The Evil Twins', Aztec Viking Pirate aka Fluff, Number 5, Rocket and Panda

OTHER PETS: A Darwin carpet python, Mitch; two cats, Hemlock and Hazel; a quail, Quailor Swift; and a Pineapple Conure parrot, Albie (abbreviated from his inherited name, Little Bastard to LB to Albie) – 'He *is* a little bastard, too. But I love him.'

SNACKMAKER

'I'm not good at mindfulness, so being able to come home and sit quietly, just watch the chickens be chickens, was a huge part of my mental health.'

Adele Scott's tattoos in order of inking. The first, the date of her marriage to husband Xander in Roman numerals: IX-III-XII. The second, a play on words from *Star Wars*: 'May the fourth be with you', written in High Elvish. 'The joke is also that High Elvish doesn't have words for "May the fourth", so it's spelled phonetically. Which means you'd be reading English, in High Elvish. The dream is that, one day, someone will be able to read it.' The third, from *Harry Potter and the Prisoner of Azkaban*: 'I solemnly swear that I am up to no good.' The fourth, an arrow from ankle to hip, a poem from Tolkien's *The Fellowship of the Ring*: 'The road goes ever on and on, down from the door where it began. Now far ahead the road has gone, and I must follow, if I can.' The fifth, homegrown produce from the garden, framing Adele's favourite chicken, Aztec Viking Pirate. The sixth, 'just because', a crescent moon. The seventh, on her chest, a Scottish dagger with a stag head. 'My brother has the same. It's a tribute to our family crest.' Last, finely drawn two months after the dawn of the fox, a row of airy chicken feathers commemorating Adele's lost girls, lying as light as, well, feathers on her inner arm. A breath could blow them away.

'Xander was going to feed the chickens that morning, because I had to be at work at 7 am. But their run was next to the bathroom and I stuck my head out of the window to say hello. One of them was lying on the ground, and for a second I thought, "What a silly chicken, sleeping in the run." And then I understood. We ran out and found their little bodies still soft and warm. I couldn't function, not for a whole day. Two days later I had a tachycardia episode where my heart beat at triple speed for ten minutes, then stopped for a moment. The doctor ran a battery of tests and I'm as healthy as an ox, so the consensus was it was brought on by grief.'

•••

Adele and Xander, freelance arts creatives, established their bohemian paradise in Melbourne's unfashionable outer west five years ago, skidding like Indiana Jones into a 'renovator's dream' by the skin of a deposit, moments before the median house price rocketed. Adele's an interior designer with many frocks and a lush eye for deep colour and vintage, who has a *Star Wars*-inspired burlesque act. →

Xander is the artistic director of a cabaret venue, Melbourne's iconic Butterfly Club. They are animal lovers and nerds who name their critters, plants and kitchen appliances for the fun (Yurtle the spurtle is their Scottish implement for stirring porridge, Booj is a food processor, Figaro and Philomena are fig trees). They're both whizzes at reclaiming found objects (their enviable copper-top kitchen bench was a manky discovery in the bowels of someone's shed), and are elbow deep in sustainability, abundance, zero waste and self-sufficiency. The kitchen is crammed with jams and home-made ice-cream and chutneys and bottled soups and kombucha. The garden, on settlement an arid rectangle, is now a seasonal riot of rosehips, raspberries, limes, pomegranates, plums, peaches, figs, apricots, eggplant, pumpkin, squash, corn and beans. There is spirited self-expression here, out the wazoo.

'Waste not' is their motto, also 'beauty', and there isn't an unproductive, unglamorous centimetre in their eclectic house, garden or

philosophy. Xander's bent is a commitment to permaculture, frugality and self-sustainability. For Adele it's a true fit with the paganism that she reconnected with as an adult.

'I discovered paganism as a young teenager, twelve or thirteen, reading a lot of medieval fantasy books. It's not about finding a deity, it's the cycle of the seasons and having respect for them. I like the rituals involved in their practices. We have a winter solstice party every year: fire, friends and food. You write down something you want to let go of, cast it in the fire, then jump over it. Pagan rituals are about community and sharing food from that season, not about having "dominion" over the natural world. It makes more sense to me than Western religions.'

Being in lockstep with the rhythms of life allowed Adele to cherish her chickens, and now to manage their loss. She doted on their individual personalities (Spike was 'a pompom with a beak held together by rage'), posed with them for Insta, coddled them, bonded with them, loved them. →

Aztec Viking Pirate suffered from pendulous crop, an often fatal condition in which a chook's crop overextends, sags and cannot empty properly or spring back. Adele went through a series of 'bra' designs for Aztec, and was rewarded by her favourite gal finally coming good. Aztec laid for the first time in eight months two days before the fox came.

'I'm not interested in having children, so they were my babies too. I nurture animals and release the mother part of me that way. I knew, though, that when you have this many pets, and attach, you open your heart more to heartbreak. But when a tree loses a branch it gets a scar, and grows another branch. I'm a growing person. Growth comes with scars.'

The garden is not the same without the animal noises. The chooks chatting. Adele misses getting up to tend the chickens, the time she spent outdoors. 'I'm not going outside as much.' Like all freelance creatives she works several jobs with no security. It's stressful. 'I think my entire generation has an anxiety disorder. Having the girls helped a lot with that. Ah, they were a huge comfort. I'm not good at mindfulness, so being able to come home and sit quietly, just watch the chickens be chickens, was a huge part of my mental health. There's something lacking. Something missing. That calmness. It's harder to get out of bed in the morning without the chookies. But you plough on as best you can. I have a quail to look after, an old man with arthritis.'

Adele's current project is a memorial to her grandmother in the front garden: a Granny Smith tree, and some of her grandmother's favourite flowers, alstroemeria, sea lavender, daphne and salvia. 'I use the garden to create meaning. Establishing roots is important. Some people see having a mortgage as a trap, and we do struggle financially, but having this place which is mine, which I can do what I want with, is the epitome of freedom. A garden that nurtures us is how to have a life.'

There will be more chickens – Adele is amused by people asking her 'if' she'll get more – but not yet. The timing must be right. Breathing space. 'You don't want to rush. You make bad choices.' Her mum replaced a much-loved labrador by adopting a couple of lab puppies. 'They're nothing like the other dog, they're little shitheads.' Another relative did the same with a German shepherd. 'Completely different personality. You have to grieve, and wait until you're ready. They're relationships.'

There must also be a fox-proof mansion. Nothing happens until they build that. Something grand, stylish, upcycled, buried deep, impenetrable. Adele's after a 'Palais de Poulet'; Xander's more inclined towards a Spanish 'Casa del Pollo'. The name game has begun, and Adele is leaning into spring, because they want chicks to raise by hand.

•••

On the evening of the attack, Xander and Adele dug five little graves. 'I took the bra off Aztec and buried it with her. We buried the twins together. We buried Spike alone, which is what she preferred. We put flowers in the graves, and rosemary, and planted apple and pear trees over them. Aztec has a champion quince. Something over them that's going to be there for years, and produce. And then we got very drunk.'

Adele looks down at her arm. 'I was still grieving when I got the tattoo. I wanted to have them with me all the time. It's a comfort, to look down and go, "Ah, there you are." But I like that it represents my love of chickens in general. I didn't want an RIP tattoo. It's a tribute to all my future chickens as well.' ●

'I'm not interested in having children, so they were my babies too. I nurture animals and release the mother part of me that way.'

DYNASTY

MEGG AND JESSAMY MILLER

Megg and Jessamy Miller, mother and daughter, editor and assistant editor of *Australasian Poultry magazine*

Live: Megg in Euroa, Victoria; Jessamy in Northcote, Melbourne, Victoria

Megg pictured with Peaches and Creamy, Pekins; Jessamy pictured with Lulu, a Blue Langshan bantam, and Spangles, a Silver Spangled Hamburg

'I've always had an ability to calm a chook. It's just intrinsic in me. My grandmother had a beautiful spirit with animals so I think that inspired me.' Megg

NUMBER OF CHICKENS: Megg: thirty-nine (thirty-eight hens, one rooster – chicken maths!); Jessamy: five (hens)

BREEDS: Megg: Anconas, Dorkings, Langshans, Minorcas, Welsummers, utility White Leghorns; Jessamy: Blue Langshan bantams, Pekins, Silver Spangled Hamburg

NAMES: Megg: Miss Neil, a Langshan who has laid 935 eggs in eight years, and Dorky the Dorking; Jessamy: Peaches, Creamy, Spangles, Lulu, Spare Lulu

OTHER PETS: A cat, Trevor

Megg Miller, editor of *Australasian Poultry* magazine since she founded it in 1990, sharp as a beak, recipient of an OAM in 2019 for services to poultry, is cranky about the internet. Not in a 'get off my lawn' way - Megg's intensely future-forward for someone 'of a respectable age' whose library of antiquarian poultry books dates back to the 1600s - but because the internet is crammed with outdated, regurgitated and recycled information about chooks. Colour her frustrated. She leans forward intently. 'It's re-running information which is fifty years old and no longer appropriate. Like, "Oh, you want a good layer? Get an Australorp." No! They can't lay to save themselves today.' This is controversial. Everyone knows Australorps are great layers. It's, it's . . . what I've read on the internet.

Megg's daughter, Jessamy Miller, the assistant editor of *Australasian Poultry* (for twenty years), editor of *Grass Roots* magazine (for twenty years), and long-time contributor to ABC's *Organic Gardener* magazine, nods her head solemnly. They've discussed this many times. 'It was the utility Australorp which laid well, and that's a rare breed now. The show Australorp eats a lot, lays a small egg, and isn't a prolific layer.' Mind blown. There's more. The Rhode Island Red? Exhibition White Leghorns? 'Cannot lay to save themselves!' There are as many opinions about chickens as there are chicken people. At this very moment someone is yelling 'AUSTRALORPS ARE INCREDIBLE LAYERS!' at the clouds. But Megg and Jessamy Miller, the affectionate mother and daughter tag-team poultry brains trust, are a fowl-focused edition of *Mythbusters*, complete with banter. For every deep-dive into the science or history of poultry, there is a shot of affectionate whimsy. These two are close, and a solid chook-based comedy double-act.

Jessamy: 'I bought the Hamburgs because they're supermodel stunning, and - you do know you grow to look like your chickens - my hope was I'd grow to look like them. But the Pekins found me and, at the end of the day, I'm a bit of a fluffy Pekin.' Megg: 'I worry about this. The turkey wattle, the wrinkles, the hair. The fat rolls and the beard. Oh, I'm definitely a turkey. But that's okay. Turkeys are beautiful birds.' →

'I bought the Hamburgs because they're supermodel stunning, and – you do know you grow to look like your chickens – my hope was I'd grow to look like them. But the Pekins found me and, at the end of the day, I'm a bit of a fluffy Pekin.'

But wit and warmth help their messages cut through, and their knowledge base is vast. They've both been immersed in chooks since birth, and between them they've got over a hundred years of experience. Of all the poultry-obsessed opinion holders in Australia, Megg and Jessamy are the gold standard of primary and evidence-based research. *Australasian Poultry* is not a frivolous read. It's up-to-the-minute in terms of breakthroughs, but decidedly old-school in tone, servicing Megg's world of exhibition poultry enthusiasts and breeders. It's practical, with an emphasis on tradition, with spotlights on breeds, profiles of poultry club heavyweights and up-and-coming juniors, tips on how to handle infestations of parasites, club show reports, and features on incubation. The photographs of the chickens are lit better than the photographs of the people. It's a niche, technical read, and their subscribers would not tolerate anything less than objective rigour. 'We try and give people spot on, up-to-date, scientifically-based, topnotch information,' says Jessamy. 'If someone says to me something about chickens, "They do that" or "I do this", I want to know where their evidence is for that, and I'd prefer it to be scientifically proven and not a single data point.'

The chicken world is essentially siloed, with backyarders rarely intersecting with collectors and exhibitors, but Jessamy and Megg have broken ground, working together to cover multiple beats and perspectives. 'We've divided the poultry world up between us,' says Jessamy, co-author of the Organic Gardener book *Happy Hens*. 'We're completely complementary. It's a big bond between us because we just talk about our chooks back and forth all the time.' Jessamy's a 'backyard chicken, kids, and what works in your garden' specialist, with five classic pet bantams in her small inner-city Melbourne backyard. 'They're my babies. I love them.' Megg's area is poultry showing, rare breeds and large flocks; she's an exhibitor, educator and judge described by leading poultry administrator Neil Penny as 'the best-known poultry personality in Australia'. Megg still runs several breeds at her Euroa property, pioneered the inclusion of eggs as a category at poultry shows, and is in the midst of a two-year experiment to learn which breeds will best cope with climate change. Future-forward, remember. 'I don't like seeing my girls die of the heat. People think they want Light Sussex, but good luck keeping any of the heavy feathers alive as temperatures in the forties become normal.' Megg's a sought-after judge, a patron of six poultry clubs and a founding member of the Australian Poultry Standards Committee, and was also a founding member of the Rare Breeds Trust of Australia in 1991.

'She just about single-handedly saved some of the breeds,' Jessamy says with pride. 'We started out with a flock of mixed chooks, and then a new breed arrived every year. Indian Games, then Anconas, then Brown Leghorns. Megg became obsessed with rare heritage breeds and put them back on the agenda. No one was keeping them, they were hard to find, they were being forgotten about.'

Megg and Jessamy's definition of family includes chickens. It always has. Unusually for lineage in the poultry world, which has had a masculine bent, the passion was passed down the matriarchal line. Megg's grandmother had a mixed farm which sold honey, eggs, milk and fruit. 'It was a dream farm. She kept over one hundred Australorps, which is a lot of chooks. Oh look, I just fell in love with chooks and farm animals as a little girl. I adore them, to be honest.' As an adult she moved onto a country property in Murchison, when Jessamy – an only child – was 'a nipper', and got straight down to chickens. It's all Jessamy has ever known. 'I used to take my ball out to try and play soccer, and the ball would roll along the ground, and by the time I got to it, every hexagon would have a poop stuck on it. And I'd be thinking "hmm". You know, this is just out in the paddock and there's really nowhere to play where there are not some chooks.' Megg acknowledges there were a lot of fowl. 'At the height of it, look, I had fourteen types of chook. All colours of turkeys, a hundred guinea fowl and seventy geese.'

They share a feature film's worth of hilarious, intense, bonding-over-poultry memories. Think Judy Davis as Megg, Toni Collette as Jessamy, maybe Baz Lurhmann directing. An abundance of vivid scenes. Jessamy crammed in a van with Megg and multiple turkeys on road trips to the royal shows in Melbourne, Sydney and Adelaide. 'So much poo. But she won Champion Turkey for like twenty years in a row.' Jessamy earning fifty cents a week pocket money for raking out the chook pens. →

'You did a good job,' observes Megg, 'you were very rakey.' Jessamy devotedly assisted as Megg tended the chickens. 'I used to walk around every night and help her feed the chooks for an hour with a wheelbarrow. And when Megg wasn't there and my dad looked after the place, I walked around with him and told him all the things he was doing wrong. "Megg doesn't do that. Megg doesn't do it like that. Megg . . ." and he'd go, "Shut up. Shut up. You're not helping. I'm not Megg!"' And when the marriage broke down, it was the chickens that pulled Megg out of her misery. 'I suppose it was depression I felt after my marriage broke up. What really made me contribute to life again was, "Oh, I have to feed the bloody chooks every day."'

Chickens have defined their mother-daughter relationship - a free-range jaunt from the get-go. Jessamy grew up with chickens and farm animals as siblings. Megg always said, 'If you're an animal person, your animals are like people to you and you love them.' Jessamy took this on board. 'I grew up thinking they were like my brothers and sisters. I'd go into the dog kennel and put the dog chain around my neck and pretend I was a dog, and Megg

'I could never bring up children without chooks. Because a childhood without chooks is not a childhood.'

would be like, "Come out of there now. There are people coming. We don't want people to see this."' Megg is still, quietly, mortified. 'How we never got reported for chaining up the child . . .'

They laugh and lean in to each other. Good memories. They're each holding a pair of Jessamy's tame fat hens, and Jessamy, being a backyarder, cuddles hers close. Megg, from the world of animal showing, automatically goes for a show hold, supporting the hen by threading her hands through the legs from beneath. But it doesn't matter how you do it, so long as you're holding a chook.

Everything comes back to chickens, and Jessamy is pleased her son Tayo is emerging as the next generation to love the birds. Chickens, after all, are everything. 'It just isn't a life for me if it's a life without chooks. I had a break in my twenties, when I was busy going to parties and having a good time, then as soon as I knew I wanted a family, back they came. I could never bring up children without chooks. Because a childhood without chooks is not a childhood.'

And that may be a single data point, but it's a fact. ●

RARE BIRD

IAN NASH

Horticulturalist, poultry breeder, champion for the fancy

Lives: Teesdale, Victoria

Pictured with his Phoenix rooster

'I look out for the hen-pecked chickens. I was bullied a bit at tech school, and I don't want that environment for my birds.'

NUMBER OF CHICKENS:
Three hundred ('You're not supposed to ask that. Would you ask a lady her age?')

BREEDS:
Golden Pencilled Hamburg, Golden Spangled Hamburg, Old English Game (full size and bantam), Phoenix, Plymouth Rock, Silkie and Silver Spangled Hamburg

NAMES:
'I have too many to have personal relationships with my chooks. But I love them all.'

OTHER PETS:
Four Jack Russells, Meg, Jock, Angus and Little Sally

A chicken's eyes, frankly, are superior. Chooks have tetrachromatic vision, which means they see in four wavelengths: blue, red, green and ultraviolet. They have a 300-degree field of vision, and their motion-sensing ability poops on ours, which explains why they can always scoot their fluffy butts just out of a yearning hand's reach. They have three eyelids, and their eyes operate independently: their right eye is near-sighted, to see food and flock members; their left is far-sighted, to watch for predators, such as birds of prey or, you know, us. Chooks keep an actual eye out for danger and scratch for noms at the same time. Impressive.

'The chickens, they're good to help with the unwinding. The calming. To get my mind off things.'

Ian Nash, horticulturalist and proud rare-chicken breeder and poultry judge, also finds himself employing two simultaneous modes of vision: he looks back and he looks forward. He's a pragmatic nostalgic, honouring the old ways but realistic about the future. His business, for example, is returning vintage gardens from grand nineteenth-century Victorian mansions to their former glory, using drought-resistant plants such as natives and succulents. 'They've got to be easier to look after. Back in those days they'd have a couple of full-time gardeners. And you can topiary and box natives; it doesn't have to be English box. There are alternatives to roses. Things have changed. Life's changed. Climate's changed. You've got to work with her; she'll have the last say.'

Ian, now in his fifties, has been showing poultry since he was a junior with the Essendon Poultry Club in the 1980s. He was from the last generation of juniors, the final swell of the golden era of Australian exhibition poultry, when regional poultry clubs were thrumming social hubs, poultry shows were galas, and there was always a new wave of young chicken enthusiasts wanting to learn from the master breeders.

'Essendon used to be the single biggest chook show in Australia – two thousand birds for a one-day show. It's dwindled to eight hundred now. Someone like me has to get my head around that it's not the same as it was in the seventies and eighties. The old breeders who used to breed amazing birds have passed, and there's not many people who have that sort of history anymore. I find it really sad. I'm lucky, though: I've got the memories of what it was like to be a junior back then.' →

Ian is now a judge, a lifetime member of the club, and a senior himself. It's his passion and his community (you don't get an Old English Game fowl rooster tattooed on your shoulderblade willy-nilly). He jokes he's 'one of the young ones these days', and has taken up the gauntlet of rejuvenation. Of attracting new blood to the fancy. He calls it 'fighting the good fight', and is framing the challenges facing the poultry clubs as a transition, not an ending. The clubs are up against cultural shift, and they need to get agile.

'Some clubs are booming, others just sit on their arse. It's a dying hobby in the city: blocks are too small and you can't keep roosters. People are too busy. Kids get bored easy these days, too, I think. It's hard to keep their concentration. They're on their devices; the games they play are so quick. I used to sit in the pens and watch the chickens for an hour.'

A 'junior' is an exhibitor or breeder under sixteen, and they *are* coming through, here and there. Just not in huge numbers. Ian says the key lies in changing the old ways of doing things. 'There's keeping the traditions alive, but the clubs have to change or they're gone. We've got women coming in, younger people – the more the better, as far as I'm concerned.' Nash agists roosters on his two-acre property for would-be breeders who can't keep males, and he actively supports new talent. 'There's a few of us doing it. If someone's keen on poultry and wants to show them, I'll put them in contact with breeders, send them in the right direction.'

Besides working on his own dry-climate garden, Briar Hill, a local landmark in Geelong, and stately home estates throughout Western Victoria, such as Murdeduke at Winchelsea, and Darriwill Homestead in Bannockburn, Nash also introduces kids to fresh veggies and chickens at Doncaster Primary School. The school has a flock of about ten birds: Silver Laced Wyandottes, Buff Cochins, Langshans and Pekins. The Year 6 students have a chicken-care roster, and Ian teaches basic chook husbandry – how to handle the chickens kindly and carefully – and when the Essendon show comes around each July, the keen beans take the birds and exhibit them. →

man
of
STIHL

'We might pick up a couple of juniors.' Another part of Nash's agenda is ensuring the birds are pure breeds, rather than hybrids. 'I prefer to push the pure-breed line. They're not scary, they're easy to pick up, and friendly. I'll get them an Araucana and a Marans soon, so they can get the coloured eggs. Then they'll see there aren't just brown or white eggs. It makes it more interesting and fun.'

Ian's mission is not entirely about the challenge of breeding the perfect bird. The greater advantage to running poultry is its benefit to mental health. Ian has a sense of urgency around this; he's seen the dividends. 'I have a couple of kids at school who don't have a stable background at home, who suffer from depression and anxiety. They help me out in the chook pen, and it keeps their minds off what's happening outside. Relaxes them. Last week a prep kid fell over during sport and started crying. The teacher took him and his friend to the chooks, and he stopped screaming in a minute.

'I suffer from depression, and I have for a long time. My mum suffered badly so it's probably hereditary. I'm on meds, but the chickens, they're good to help with the unwinding. The calming. To get my mind off things. When Mum was in a home in Ascot Vale, I got a cage made up with bantams in it, and would move it around outside different windows. The men and ladies would sit and watch the chickens, talk about them. Gave them something to look forward to each day. That's important when you've lost everything.'

Nash sees himself as one of the rare birds he breeds. The proud, upright, Old English Game bird on his shoulderblade, for example, is no longer a sought-after chicken. 'They're surplus to requirement these days. Not for everyone. But it represents my passion. I enjoy the rare side, to bench a good bird that's rare and put it up for a major award. That's satisfying.' ●

CONSIDER THE ROOSTER

THE PLIGHT OF MR FABULOUS

You will notice, in Chinese Astrology, that there is a Year of the Rooster. Not a Year of the Chicken, or of the Hen, or of the Chook. None of the other eleven zodiac signs (pig, horse, dragon, tiger, doggo, etc.) are gendered. The ancient Chinese wanted themselves some *male* chicken personality traits, and if born into that year, you are considered to be proud, vain, confident, pompous, attention-seeking, flamboyant, hard-working and brave. Rooster celebrities include Beyoncé, Britney Spears, Ariana Grande, Peter Dinklage, Honey Boo Boo and Tony Abbott. The traits are also a fair portrait of my dad, Arthur, who was born in 1921, the year of the metal rooster, and wore a powder-blue safari suit during my formative years and wasn't afraid of loudly sharing his achievements and opinions. Roosters are not backward in coming forward.

Traits that land on people as back-handed compliments sit nicely, however, on the male *Gallus gallus domesticus*. There is scarcely a more glorious creature than the rooster. Charismatic, ostentatious and pulsing with chicken testosterone, a cock in full strut puts any hen to shame in terms of sheer pimping plumage and allure. Roosters ooze *je ne sais quoi*. Inexperienced chicken buyers inevitably return from the market with at least one roo, because the boys are the best looking. If you're drawn, inexorably, towards a particularly fine-looking chook, intoxicated with its pizzazz and personality, trust me, it's a rooster and your heart is about to break.

•••

This is the thing. You cannot keep your rooster. Many of us wish to – besides being spectacular, they are of value to the flock and are often very sweet and courageous – but unless we live out country, we generally cannot. The most intractable aspect of backyard chicken husbandry is what to do with the menfolk. They are, shall we say, incompatible with suburban living. →

'If you're drawn,
inexorably, towards a
particularly fine-looking
chook, intoxicated
with its pizzazz and
personality, trust me,
it's a rooster.'

Some neighbours are willing to be bribed with fresh eggs or a placating jam sponge but, on the whole, inner-city types are a) weirded out to be handed a random cake, and b) not enchanted at being woken at 4.30 am by Mr Fabulous. Not that your council even allows roosters. No, they do not.

There are ways around that. Councils are mostly concerned with noise, so if you control that, they tend to care a whole lot less. People get creative. There are 'crow' collars, which stop your rooster from fully extending his neck: their success is mixed, at best producing a strangled rather than full-bodied crow. Ditto for building your rooster's perch close to the roof of the coop: if he can't stretch up fully, the theory goes, he can't quite do the do. Determined rooster fans put their boys to bed in dark boxes and shut them in the garage or shed. This stops the lads from glimpsing the dawn, and delays the crow for a half-hour or so. Roosters have superb circadian clocks and don't need to see first light to know it's coming. Actually, due to their red junglefowl ancestors, they'll start crowing up to two hours before dawn, in what's known as 'anticipatory pre-dawn crowing'. So, yeah, fair call, that would wear thin pretty quickly. Essentially, what you're talking here is soundproofing. A rooster in a box in a garage probably won't disturb the neighbours. It's labour intensive, in contravention of most councils' regulations, but works for an increasing number of roo lovers. For those who wish to retain a rooster or two, the balancing act is worth it. For most city slickers, though, our love must remain unrequited.

At my place, one clutch of Rosecomb chicks brought forth nothing but male children, and of course you get attached and I made a bid to keep one: young Percy. Percival felt like a grand rooster name, but this chap was small and fat and faintly ridiculous, so Percy it was. I figured I'd give it a go, and if the neighbours complained, off to the country he'd go. A couple of months went by, and I began to hope. As a bantam his crow wasn't *so* loud. Cute, even. Maybe the flats next door *liked* a touch of the country in the city?

They did not. I came home to a card, slid under the front door, from my local council, requiring me to call on the subject of 'poultry breeding in domestic premises'. I immediately felt sick. Fark. I could lose all my chickens. I'd put myself on their radar. What an idiot. The council poultry wrangler, as it transpired, was a lovely man and totally pro-chook. He did not want to take my chickens away. But he did read me the time log from my (unspecified and clearly tortured) neighbour, who'd kept a 'crow diary' for several weeks, detailing their experience in short, painful increments. 'February 17th: crows at 5.08 am, 5.14 am, 5.27 am, 5.35 am' and so on and so forth. *Mate*. So sorry. Percy, the adorable little blighter, had to go. Apart from anything else, our garage had no doors.

Percy was lucky. He had somewhere to go. My friend Nicholas (the Rosecomb enthusiast) was willing to take all our roos. He has an eye on breeding, and Rosecombs, a tiny ornamental show bird, are not common. But it is unusual to have an easy solution. Australia is, essentially, drowning in excess and unwanted roosters. More and more of us are discovering the delight of having chooks, and the joy of hatching our own eggs. More and more schools have chicken-hatching programs. We are in the midst of a chicken renaissance! And all of us wait with bated breath after a hatch, crossing our fingers that the particularly fine chicks we've got our eye on don't develop saddle feathers or an engorged comb. It's awful. Not because we *don't want* them, but because we *can't have* them. The reservoir of unkeepable males is full to overflowing, and where do they go? Even country chicken enthusiasts only have need of, or room for, so many. Roos end up dumped illegally; rescue facilities are bursting with them, and they commonly end up culled and in someone's bin or freezer. If you eat chicken, culling your own young roosters is the optimum way to do it, but not everyone's up for sharpening the axe and just want the problem to go away. →

The two chicken subcultures in Australia that wholeheartedly welcome and value the roo are breeders who sell pullets to people like you and me, and the show community. But even for breeding, the required ratio of cock to hens is somewhere in the region of one to ten, as roosters prefer a harem to one special lady, and can get fighty if there's competition. The show community are passionate about beautiful roosters, and poultry shows teem with stunning, ribbon-winning, strutting males. But, again, you only need a handful of top-notch roosters to propagate your line, and excess birds eat a lot of feed, so breeders cull hard, cull often, and make a lot of chicken soup.

In the world of commercial egg-laying, of course, it's even less fun for Mr Fabulous. That much is obvious when you think about the scale of the egg industry, and then combine it with the common-sense revelation that for every hen hatched there's an unwanted cock pipping his way through the shell next to her. Every year in Australia 12 million one-day-old male ISA Browns and Hy-Line Browns are fed into industrial grinders, because male layers are commercially worthless. Can't lay eggs, bro. So they're macerated instead, which is the industry term for 'shredded alive', which doesn't bear thinking about too much and surely makes hatcheries emotionally complicated workplaces. *Cheep cheep CHEEP!* It's instant, so arguably kinder than the gals' lot, which is eighteen months of pumping out eggs followed by a trip to the abattoir once they're 'spent', but, hey ho, from a chicken's perspective I suspect it's a Sophie's choice.

It's an awkward realisation, for those of us who are keen to Do The Right Thing, that free-range and free-range organic layer chickens come from exactly the same hatcheries as caged birds. There are no 'guilt-free' eggs, in that the boys are always wastage. The industry is finally getting onto this, because of course it's a) bad optics and b) cheaper to not have to incubate 'male' eggs for twenty-one days. (For a sense of scale, that's 6 billion unwanted male layer chicks macerated per year worldwide.) 'No kill' eggs went on sale in Berlin at the end of 2018. Biotech researchers prompted by the German supermarket company Rewe Group discovered a method of determining the sex of a chicken foetus on day nine of its incubation. They take a fluid sample from each egg to detect a chemical biomarker present in female chickens. They've called it 'Seleggt', and I can't decide if that's terrible or genius. Meanwhile, in Australia, scientists at CSIRO are using gene splicing CRISPR technology to paste a gene from a sea anemone into the male-only chicken chromosome. That gene produces RFP, or 'red fluorescent protein', which glows from within the male egg under laser light. This can be picked up directly after fertilisation, so incubation is terminated. A win, of sorts, for beleaguered roosters, which have truly not enjoyed many benefits from late-stage capitalism.

...

Roosters, the poor magnificent bastards, have become surplus to requirement. This is a recent development. In the beginning – if we go back thousands of years – it wasn't meat or eggs that first made chooks hot property, but the roosters' propensity for fighting (considered perfect for gambling and entertainment), and their insistence on waking the world slightly pre-dawn. It's only of late that we've all had an alarm clock or smart phone. Or even banned cockfighting. Among all animals, roosters were heralded as unique and necessary. →

In Greek mythology cocks were dedicated to Apollo, god of light, because they gave notice of the rising sun. They were also dedicated to Mercury, god of shopkeepers and merchants, because they summoned men to work. And to Aesculapius, the Greco-Roman god of medicine, because 'early to bed and early to rise makes a man healthy' etc. A weathercock – a weathervane in the shape of a rooster – was set up on every church steeple in the ninth century by papal enactment. It was the emblem of St Peter, a reminder that he denied Jesus three times before the cock crew twice. In Muslim legend, the chief angel in the first heaven is an enormous celestial rooster whose crest touches the second heaven. He is favoured by God, and his crowing wakes every living creature on Earth except man. On the day it ceases to crow, the day of judgement is at hand. Muslims are exhorted to not curse the rooster, because he wakes them for prayer.

Roosters, I'm saying, used to be a big deal. Now it takes a Willy Wonka golden ticket for a lucky bird to live and spread his seed. We've lost the cultural agreement that chickens are part of the village. We forgot a lot when we took them out of our gardens and consigned them to factories. We no longer consider a rooster crowing as natural and welcome, or think that roosters are useful and necessary. We've forgotten that a flock is better with a cock.

The problem for roosters is that flocks operate fine without a rooster, at least from a human perspective. The hens get on with it. The pecking order gets a bit meaner. Oh well. Rooster tasks get divvied up. A dominant hen might start crowing, and she'll also watch out for predators, as will the ever anxious chicken at the bottom of the pecking order. It's all a bit Rosie the Riveter. In my flock, for example, Violet is currently crowing – a polite lady crow, but it's happening – and Lally's been, ahem, mounting Blondie. Again, common behaviour when there's no roo.

But flocks function more comfortably, more fluidly, with a good rooster. He patrols for danger and will protect his hens, often with his life. There are countless stories of courageous roosters who died trying to defeat a fox, a hawk, a dog. Some of them even win. They 'tidbit' for the girls, finding treats for them, and again, if they're good roosters, they'll let them eat first. They flirt and do come-hither dances with their wings dusting the ground. They are romantic and gallant and pompous and funny. They'll make baby chickens for you. They are often affectionate, particularly if you're female and a rooster decides you're his hen. And there is something peaceful about watching a flock move as one, with the rooster calling the shots. Roosters have a job, and they do it very well.

No, they're not all charmers. Their instinct to protect can translate into aggression, although *theoretically* they shouldn't view the bringer of food as a threat. My grandparents had chooks, and we had to carry an extra pail at feeding time because Peter, the rooster, was an arsehole. If you didn't put the bucket on his head as soon as you opened the door, he'd lacerate your legs. Peter went for Dad once too often, and turned up roasted one Sunday dinner. He had a diminished fan base at this point, and from memory the most concerned comment was 'pass the potatoes'.

I miss Percy, though. I liked the crowing. And having had a rooster, I prefer it. Something clicked into place. There's no question that, although chickens are supremely adaptable, the closer the flock is to balanced, the better it functions. Hens, after all, don't live to lay eggs, they live to become mothers. You've not seen happiness and pride until you've seen a chook hatch some chicks and, biology being what it is, you need a rooster for that. It wasn't that long ago, a handful of decades, when chicken sounds – crowing too – were just a regular part of life's soundscape, no more noticeable than today's background thrum of traffic, which, let's face it, we've internalised and tolerate without comment. Every time I've moved house, for the first two or three days I've woken in a sweat due to the appalling traffic noise. Within a week my brain flags it as normal, and I sleep right through. Perhaps, as our love affair with chooks rekindles, we can reincorporate the idea of a rooster crowing being common or garden. Splendid, even. And we can welcome them back. ●

CHICKEN MILLINER

I'd love a rooster. I think chickens need a rooster. They're in paradise here but they don't have a man.'

NUMBER OF CHICKENS:
Six (hens)

BREEDS:
Cuckoo Pekin, Silkie, Frizzle Pekin cross and Silkie Frizzle Polish cross

NAMES:
Joan Jett, Carmen Miranda, Blossom, Penny, Winnie and Dora

OTHER PETS:
A Maltese Shih tzu, Sunny

MANDY WATTS

Instagram and Etsy queen, rustic garden stylist, purveyor of smiles

Lives: Toowoomba, Queensland

Carmen Miranda
MAY 2013
Lucille 23 April 2015
13 Dec 2014 Bertha
Blossom
11 Sep 2016
Joan Jett
3 MARCH 2018
Megg
3 Mar 2018

You can't overstate the delight Mandy Watts brings into the world with her exquisite little crocheted hats for chickens. Adorable Brunhilda viking helmets with blonde plaits. Darling Easter bonnets with pompoms, and cloche hats with flower detailing. A mohawk (oh my god, the mohawk). A unicorn. A Yoda. A crown. Santa hats with beards. Koalas, Easter bunnies and alpacas. Swooningly cute. Intended to be popped briefly on broody hens for a photo opportunity, Mandy's creations are responsible for a profusion of Instagram posts, personalised greeting cards, and people clasping both hands to their breast and exclaiming 'Awwww!' One of Mandy's Instagram friends described her as a 'purveyor of smiles', and Mandy often finds herself laughing as she crochets something fresh.

'I even made a captain's pilot cap, with the emblem and the blue peak. That went to an actual pilot in Canada who has a special-needs chicken called Hoppy.'

The whole thing started as a laugh, when she saw another Insta friend's chook, Fatty, wearing a cowboy hat from Spotlight, and knocked up a few beanies for fun. She has more than 31,000 followers on Instagram at @herechookchook, and a bustling shop on Etsy, which ships her chook millinery worldwide: Italy, France, New Zealand. 'I sent one off to Croatia yesterday.' To date she's sold over a thousand hats, which is a lot of meditative crochet, as each piece is unique (she follows inspiration rather than patterns) and takes an hour, minimum, to make. Her biggest market, in a surprise to no one, is America, and she does a brisk trade in holiday tie-ins: Christmas tree hats, witch and pumpkin hats, 'Uncle Sam' top hats, and pilgrim hats complete with teeny crochet buckle detailing for Thanksgiving. Mandy's been sewing since she was twelve, and is a top-shelf creative.

'I've always been a maker. I can make virtually anything. People ask me, "Can you do this?" And I do. I even made a captain's pilot cap, with the emblem and the blue peak. That went to an actual pilot in Canada who has a special-needs chicken called Hoppy. I was pretty proud of that one.'

Mandy, in her fifties, has inadvertently become Australia's top social-media chicken influencer. Not what she set out to do. But there you go. Life, eh? Takes you by surprise. →

'I was posting chicken photographs on Facebook, and my daughters said they would unfriend me if I put up one more chicken photo. Then a cousin threatened to block me. And I thought, "Okay, no one wants to see my chickens. Fine. I guess I'll just make an Instagram account for myself."' Insta, it turns out, can't get enough of brilliantly composed pics of foofy bantams living their best lives, pootling about in a scenic paradise, and her account took off like a lit fuse. Mandy now has a thriving online life, documenting her girls and the constant reimagining of her garden, which suits her fine. She doesn't much like leaving the house, and a large, warm community of like-minded chicken enthusiasts comes to her. Her husband Jason, when he's not building her a coop or rustic shed, is a truck driver working twelve- to fourteen-hour days, and Mandy does appreciate the 'social' aspect of social media.

'It's like having penpals. I've made close friends. People I would never have met otherwise, both here and overseas. I could just turn up at their house, walk in, and know I'd be welcomed. We

chat every day. Not just about chickens, but our lives.' Mandy diligently spends hours each day responding to messages from followers and uploading content. 'I try, if anyone asks me a question, I try and answer, and it's hard when you have a lot of followers and a lot of questions.' There's no resentment there, just an acknowledgement of what it takes. It's essentially her job now, particularly since she stopped work to provide care for her grandchildren. She loves it, it gives her purpose, and she's made over $8000 from crocheting chicken hats. And, when there are triumphs, hatchings, deaths or dramas – and there are many – her community celebrates and commiserates together. Mandy recently lost two of her most charismatic and loved chooks, Boom Boom Bertha and Lucille, and she was overwhelmed by the support. Many hundreds of messages. It really helps, to not grieve alone.

'Jason's away, so I have to bury them myself, and that first shovelful of earth is very hard. I cry every time. You've got to be a bit hardy, I think, to have chickens as pets, because they do die all the time, you know.' →

Mandy's in an interesting space. She's using chickens as a way of healing trauma, but it's bittersweet. Sometimes leading a chicken-loving life means she needs to draw on the resilience she's developed to deal with grief. Mandy hasn't, truly, been the same since a sucker punch of family deaths around the turn of the century. First her sister Trudy died of breast cancer, then her father passed, and finally her son, Adrian, died aged eighteen in a ute accident on a family farm.

'My mum, who's seventy-eight, almost seventy-nine, will say to me on a daily basis, "Do you remember such and such?" She'll talk about the kids when they were young, and I'll say no and she'll say, "How can you forget that?" I'll say, "I don't remember." I've blocked everything. I think because of the trauma I've had, I've lost my memory.

'That saying, *time heals*? That's a lie.' She laughs. 'Look, you learn to live with it. I'm friends with a lady on Insta whose husband died at forty-something of a heart attack, so unexpectedly, and I talked to her a lot about grief. She posted the other day that she went somewhere with her son and actually had moments of living not existing. And I thought, "Yeah, that's it exactly." I think for a couple of years you exist. It takes a while to live again.'

Her daughters, Carmen and Sophie, have helped, and then her grandsons. 'Every time they were born, well, it keeps you going. It's like the hatching thing, seeing the little baby chickens: it brings out all your mothering instincts. It's beautiful.'

Mandy has put in place things which give her contentment. Her garden is constantly evolving and is film-set picturesque, an ongoing project. It's a rustic wonderland where both her daughters married, and no matter where you point your camera, there is a potential chicken-forward Instagrammable moment. There's also a 'maternity wing', where her broody chooks go to hatch.

'I love watching them be mothers. It's what they live for, I'm certain of that. To watch the mother hen walking around with their babies: they sit there and they put their wings out to have a rest and the chicks get under, and she closes her eyes and I think, "God, they're just . . . that's happiness right there."

'The night I know they're going to hatch I can't sleep, I'm so excited, I'm thinking, "Is it daylight yet?!" so I can go up and watch. I go out to that henhouse when it's first light, about quarter to five, and I sit there in my pyjamas. I get a seat and my coffee and I've got the phone, and I send pictures through to Mum 'cause she's on the other end, she's awake too. "How many have hatched, what colour are they, oh, aren't they beautiful?" We're backwards and forwards, it never gets old. Hatching new life is the antidote to death, and I am addicted to it.'

Mandy lives in the inner suburbs of Toowoomba, a no-rooster zone, and she's had to stop hatching chicks because finding homes for the roos is too challenging. 'I've had to get chickens put down just because they're boys. And they're beautiful. You know how beautiful the roosters are. And for no other reason than that they're a boy. That's not fair, is it? You can't just bring something into the world knowing that if it's gonna be a boy it has to die.'

Mandy's popularity is really not about the hats. She had 19,000 followers before the hats. As she says, she's got followers who don't give a shit about the hats. It's about radiant chickens in a utopian setting; Mandy created the garden around the needs of her girls. It's an actual chicken paradise with shade, places to perch, plants they can't kill, places to dig, grass to eat. She gets her happiness by maximising theirs, and that generosity and devotion are what you see. 'There are accounts with chickens wearing things all the time, and the chickens are in the house, and I think, "When do those chickens get to be chickens?"' Mandy would never do that. She hardly handles them. She just pops a little hat on a broody Blossom once in a while.

'I just . . . I don't know. How do you say how much you love something? They rule my life a little bit.' She laughs. 'Ah. I know what they like, and that makes me happy too. And all of this is just to make people's lives a little bit brighter.' ●

FAMILY

SHANE SECOMBE

Horticultural officer at Alice Springs Correctional Centre

Everett, a Pitjantjatjara man from Areyonga, and Matthew, a Luritja Pitjantjatjara man from Hermannsburg, low-security prisoners

Live: Alice Springs, Northern Territory

Shane pictured with Wayne, a Pekin rooster

'One of the other officers said, "It's a two-dollar program with a million-dollar outcome." You can spend millions on rehab, then you just get a couple of chickens.' Shane

NUMBER OF CHICKENS:
About thirty (roosters)

BREEDS:
Game birds, crosses of all sorts, two surrendered Pekins

NAMES:
Whei (which means 'little boy'), Wayne, Frank, Mr Wesley

It's a brand-new blue-sky day in Alice Springs, and as the sun rises over the horizon and hits the trees it casts long bars of shadow across the red earth. Skinny fingers slicing a geometry on the ground. A tall, grey-haired, outdoorsy prison officer with a lightly toasted pink face crouches down inside the chook compound and calls out, 'Here family! Here family!' and thirty or so surrendered roosters belt up yelling hello and eager for grain. The second-chance roosters of Alice Springs jail. 'It's my constant. They're just happy to see me, whether I'm in a bad mood, good mood, the world's fallen in . . . I just sing out "family" – and they'll come.' He loves those boys.

The story goes – and with Shane Secombe, horticultural officer at Alice Springs Correctional Centre and yarner-in-chief, there's always a story – that years ago one of the many low-security male prisoners who'd worked outdoors with Shane said he was lonely. He had no family, no countrymen. Shane, who left Bendigo in Victoria for a life in Alice, and is the only officer to work solo, gestured around the grounds to indicate his own isolation. To establish some connection. 'I've got no officers and no Victorians. No family.' And the lonely man pointed to a big white goose and the flock of roosters and said, '*That's* your family.' Now it's part of Shane's lore. 'It just became, well, they're my family. It was an off-the-cuff remark and everyone picked it up. Suddenly it was "Mr Secombe and the big white goose", "Mr Secombe and the chooks", "The family." Now I just sing out "family" and up they come.' Shane, the sole architect of the prison's horticultural arm, has been giving the roosters of Alice Springs a home and new life since he joined the jail when it opened in 1996. He's outlasted nearly everyone, and unironically calls himself 'the old man', but at the prison, in Alice itself and even with his wife and children, everyone knows Shane is 'the chook guy'.

'I reckon half my calls are about chooks, not work-related stuff,' says Shane, patting his phone. 'I'm in trouble if I ever get audited. The staff, the officers, they'll say, "You're getting a lot of calls from people," and I say, "Well, it's about a chook."' In Alice, as with everywhere urban, people have roosters, and a familiar scenario unfolds: they love their roosters, the roosters piss off the neighbours, and the council has to find a solution. 'Chook people are chook people. People are attached. Even if their rooster is being a pain in the bum.' Alice Springs isn't most towns though. Remote and enclosed on all sides by desert, it has a hothouse vibe; everyone knows everyone and each other's business. The phone number of 'the chook guy at the jail' is currency, passed on at every pub and barbecue in town. Shane's evolved into a community service. 'It's not publicised, but over the years it's developed that, yeah, the jail is the place where roosters are given a chance. They're gonna come out here and have a go. It makes it easier for the rangers to have the conversation. The rooster's being sent away, not put down.'

The people of Alice are grateful, but the benefits of those scrappy rescue roosters flow both ways. The roos are central to the rehabilitation opportunities offered to low-security prisoners, and also function as a small spot of vivid kindness which – ever so slightly – softens the intense foreboding radiated by the correctional centre itself, a squat unlovely compound surrounded by building-height loops of razor wire. The prison houses around 500 Indigenous inmates, and is not, by any stretch, a 'good news story' generator. It's not in Shane's job description to humanise the jail, or make it a less shit place to be, but, on the down-low, it's what he's taken on. And the chooks are front and centre.

Shane's 9.6-hectare kingdom, adjacent to but separate from the correctional centre proper, is an oasis of irrigated banks of plants, marshalled recycled materials, a little rooster graveyard, and 'in process' gardening projects. It's peaceful, tidy and calm, has its own jaunty, hand-painted sign at the turn-off, and is a key part of the rehabilitation arm of the prison. At its centre, opposite the rambling office, is the sizeable rooster yard. It's got high wire on all sides, to protect the roos at night, but during the day they're let out to live life. 'There's enough locked up around here.'

For five days a week, fifteen or so low-security prisoners come and work each day with Shane. Women on Tuesdays and Thursdays; men on Mondays, Wednesdays and Fridays. There's physical activity, like brush-cutting and planting, and time looking after the chooks. It's a quiet place where the prisoners can decompress away from the pressure-cooker of the jail. Good chats happen when you're just feeding the chickens. It's taken Shane years of cajoling the administration to allow female prisoners access, too, and it's been a recent win.

'There are people who've been in here for fifteen years and told they're not much. Then I finally get them out here and they blossom. They're thriving. There was one woman I had, an ex-prisoner now, and she missed her kids so badly. She said, for her – she had a six-month sentence – the one thing that got her through was to come out and physically cuddle a chook. She'd come out and just cuddle a chook. She had something to mother.'

The chooks give the prisoners something to care for, look after. Somewhere to put their human needs. Shane says the women immediately go into caring mode with the chooks. The men fan out more and take on the safety role. 'They're more protective. They'll check for dingo tracks. See what's been through during the night.'

Matthew, a Luritja Pitjantjatjara man from Hermannsburg, says they come and sweep the area clean, and do a safety check. →

'Check the roosters aren't sick. Feed them. Make sure there's no snake 'ere. Make sure they're safe.' The dangers are real. The old jail, Greenbush, was in the town proper. This one's isolated in the desert, 25 kilometres out of Alice. There are predators everywhere. Matthew's cousin, Everett, a Pitjantjatjara man from Areyonga, says everything changes when they're outdoors with the roos. 'Yeah, the mind clears, open y'know. The chickens are good. Takes us outside. Inside it's lookin' at the people stressing out, talking, fighting – it's the reason we came out here to just get away from there. 'Cause someone always gotta be control the remote . . .'

The mental health benefits of chooks are now well documented, and Shane watches that magic work on prisoners and staff alike. He'll arrive of a morning and some officers will be having a meeting while they feed the chooks. Shane's been pushing for that. 'It can be very tense up there. I just thought, if I can get 'em out here with the chooks I could get that little smile. And that's the crack, the way in.' Shane's always looking for the crack. For the softening. He's consciously made being 'the chook guy' part of his identity, and in a workplace environment where emotions are routinely intense and staff under pressure, he is a beacon of non-threatening accessibility. Everyone relaxes with Shane. They tease him, even. He welcomes that, too. He likes giving the prisoners an outlet. The male prisoners made him an 'old man' chair out of wobbly branches, the women made him a jokey quandong necklace with clothes pegs on it after he had difficulty hanging some towels. He clowns around, wiggling his bum when he demonstrates how to use a brush-cutter. 'The women say they can't, so I muck around to make them laugh, then they'll try.' He works to raise the prisoners' confidence, giving them tasks that stretch and challenge them, and lands more as mentor than guard. 'He's trying to help us change our life,' observes Everett. 'He don't want to see us back here.'

Shane and the roosters release the pressure. When a new roo arrives, the prisoners compete to name it, an opportunity for playful cheek, a dig at the officers. 'They're probably more excited than me. "Look at this one, he's got fluffy hair", "What are we going to call him?", "He looks like Mr Donaldson, but don't call him Mr Donaldson otherwise he'll get the shits."' Shane acts as a release valve for the prison officers, too. He is, after all, the chicken-loving outlier hanging out at horticulture. Just by turning up, he changes the mood. 'Secombe!' they say. 'How's the chooks?'

Sometimes Shane gets a call. Something's going down in the compound, can he come? One time a man locked himself in a room, wielding a fan as a weapon, out of control. Shane was summoned. He identified himself through the door. There was a pause. 'Mr Secombe?' the man said. 'Eh, how's your chooks? How's the family?' There was the crack. The tension went out, the humanity came in, they had a chat. Shane's a circuit breaker. Wherever he goes in Alice, everybody asks him, 'How are the chooks? How's the family?'

Another story. A sad one. Shane's wife, Emma, is a midwife. She was called to a stillbirth, a young Indigenous woman whose baby had died. The hardest delivery of all. Emma was trying to connect, to help, and it was a struggle. The woman had been in the prison, so Emma thought, 'Maybe . . .' and said, 'My name's Emma Secombe, do you know my husband?' 'Ah,' the girl said, 'Your husband, he's the chook man. Wears shorts.' And there was the crack. The way in. This time to grief.

Shane's been working in prisons for twenty-five years. Last man standing. Old man. Whatever he's doing is working. 'I try and do the best. I'd like, y'know, when I'm in the coffin, for people to say, "That bloke tried. He got up in the morning and he tried." Because I care. It would be a horrible job if you didn't care. I do what I can to change an "I can't" to an "I'll try". To help people out, get a smile on someone's face.'

The flock's changed again. A young woman in town had to surrender her two Pekin roosters, Wayne and Frank. They're floofy and plump, clearly pets, utterly socialised, and must surely wonder how they've gone from cuddles on the couch to a boys' club in the desert. They are poignantly, ridiculously, out of place. Still, whenever a human appears they trot up with hope and purpose. No one at the jail has seen their like. Everett and Matthew can't stop smiling at them. 'I like those one with the feathers. They're cute,' says Everett, holding out some grain. 'They different from the other mobs.' Shane can't stop snuggling them. A big man unashamed of holding little roosters to his heart. He buries his face in their feathers. Family. ●

HORTICULTURE

TENDING THE FLOCK

JANINE ZSCHECH
Chicken sitter, poultry educator, Baptist
Lives: Sheidow Park, South Australia
'I love chickens. Chickens are good. They get up early. They're industrious!'
NUMBER OF CHICKENS:
Twenty-ish (hens)
BREEDS:
Ancona, Buff Orpington, Double-laced Blue Barnevelder, Exchequer Leghorn, French Wheaten Marans, Pekin, Rhode Island Red and Welsummer
NAMES:
Kiwi, Apple, Plum, Pear, Lavender, Annie and May, Grace and Queenie, Vegas, Dallas, Madison, Nutmeg, Cinders, Ashes, Barbara and Opal, Orange, Cheezel, Brownie, Blackie, Splash the rooster and an unnamed randy rooster
OTHER PETS:
Two rescue cats, Jim and Fizzy

Chickens get up early. First light. It's in their contract. So of a morning, every morning, be it a Monday or a Sunday or a Christmas Day (she's booked solid on holidays), Janine Zschech (pronounced 'shek') rises pre-dawn, has breakfast, lets her own chookies out, loads up the car with gumboots and bags of cleaning and anti-pest products, and goes about her rounds. Conservatively dressed, ex-corporate office, brisk and professional, Janine takes her chicken-sitting business – and biosecurity – seriously. Your chickens will not be getting sick or smothered in mites on her watch. She's at her first job by first light, pulling on the boots assigned to that house (and that house only), letting the client's chooks out, checking their feed and water, doing a quick clean, collecting eggs, spending a bit of quality time. All of the things that chickens need when their humans are on holiday. Then to the next client. And the next. And repeat at dusk for lock-up.

'I have an attachment to all my girls. I've got very good at reading them. I can see what's coming. The night before I think one will die, I tell them what a lovely girl they have been.'

Her life is shaped by sunrise and sunset, a love of service, chickens, her husband Lionel, children Austin, Tara and Psalm, and her faith. On Sundays, they worship at the South West Baptist. 'Church is always at 10 am, so I am easily home before that to get dressed. If I have a once-a-day visit on Sundays, we'll go in two cars, and I drive there straight after the service.' Janine places a high value on being useful. Needed.

Dedicated, principled and thorough, she is pushing an unashamedly pro-chicken, myth-debunking barrow. It's working for her. The Adelaide Chicken Sitting Service has about sixty clients on the books, and she can't keep up with demand. 'It's a necessary business. I love it because I'm helping. People can finally get away! They buy chickens and don't realise you can't put them in a kennel like a dog or a cat.'

There's so much people don't know. Backyard chicken owners are surging across Adelaide, often with more enthusiasm than knowledge. Janine provides answers, even giving chook advice on Sonya Feldhoff's afternoon show on ABC Radio 891. She enjoys sharing her knowledge, as well as being reminded of her heyday in the 1990s, when she had a regular show and was the Telstra Dial It Weather Service voice girl on Adelaide's Christian radio station Alta Mira (now 1079 LIFE FM). →

Janine is all about research, backing herself up, giving sensible advice. On mites and worming, cutting back on treats in spring, on how grass can cause impacted crop. Her responses are thoughtful and in-depth. Unless you're treating chickens lightly, that is. A post on her Instagram feed reads: 'Dear real estate agents, I do NOT house chickens for your clients! Consider them a selling ASSET. People fall in love with the idea of HENS.' She's indubitably correct – the wind blows pro-chicken these days – but she also has no interest in hiding someone's precious chooks away for the purpose of an open house inspection. There's biosecurity for one thing: mites and respiratory diseases are *contagious*. It's a dangerous and amateur move to mix someone else's chickens with your own. For another, there's no shame in high chicken visibility. Janine's own garden fell to her chooks a long while ago, and while it's biblically abundant with fruit trees and bushes, the kids' trampoline is now a chicken shade in summer, the potting area doubles as a quarantine, and there's no longer a pretence that the grass is suitable for human frolicking. →

'Yeah, the poo. Everyone hates the poo. You can't have a barbecue in the backyard, no. When we used to have only a few chickens, we'd go out there with the hose and blast it all, and then you wouldn't step in anything. But I now have that many chickens, you can't blast it away. At some point you just give up and walk away and go, "That's their yard."'

This is fair and reasonable prioritising from Janine's perspective. She and Lionel are essentially frustrated country people living in an Adelaide suburb, and practicality is key. 'It's a dream to move further out. We often say, "How many fruit trees can we squeeze in here?"' This from a woman who has found room for twenty-something chickens in a moderate back garden. But she's seen clients take the exact same journey. 'First up they can be absolutely prissy on the garden, and then, after a while, they let go and say, "I don't care about my garden anymore, I just want the chickens." They have an absolute change in mindset.'

Janine's done a complete 180-degree turn herself. Initially she had zero interest in getting chickens. She grew up on a mixed farm, and her one job was to look after the chooks. All five hundred of them. 'I was so sick of it. And when I finally left home to come to the city and work in an office, I swore I'd never go back to the country, and I swore I'd never go back to chickens. Ha!'

In 2011 the neighbours' chickens started flying into Janine's yard to lay eggs. And there were a couple of Silkies with curled toes who couldn't roost. Janine fretted for them. When the neighbours threw their hands up and implored Janine to take them, well, what do you do? You have to get a coop. And then you want to do it *right*, so you get some ISA Browns, and then you decide to try some other breeds and *bam*, it's all over. 'You get two chickens, and maybe it becomes three. You think, "I could do five. How did I suddenly get ten? Ten's not that many. Twenty-five? Sure, let's have twenty-five. Who's counting anymore?"' From a practical perspective Janine says that having a large, varied flock means she keeps learning – 'Problems always come up in a larger flock' – but, let's be clear, she's hooked. 'I have an attachment to all my girls. I've got very good at reading them. I can see what's coming. The night before I think one will die, I tell them what a lovely girl they have been.'

She loves the different coloured eggs, how the Buff Orpingtons sit on her lap and have a chat, how having chickens relaxes her. 'I did all the corporate stuff, I was a "super temp", in HR and recruitment, but I wouldn't go back. I'd get panic attacks, you know, thinking about going back to that life. I can't imagine it. I hadn't gone back to work after the birth of any of my children as I had lost two. That sorted my priorities out for me. I like being out. I like the freedom of getting out and roaming around with the chickens. It's like garden therapy, but chickens are even better. They talk to you, and you can do a bit of talking back, and it's a bit like . . . you know when you have laughing fits with somebody and suddenly it just brings you right in and calms you down?'

The Adelaide Chicken Sitting Service isn't just a business, it's an almost inevitable vocation. 'Who knew that the chickens would be the one thing I actually took with me? That this would be what people wanted and needed?' The business only came into being because Janine offered to come into Sheidow Park Primary to look after the school chooks on holidays. 'Lots of mums came up and were going, "I've got chickens too, and I don't know what to do. Uh, Janine, would you mind?" There was no one to do it. I went, wow. There's really a need in the community. I've gotta go do this.'

Janine and Lionel's entire ethos is contribution. It's part of their faith, and it makes them happy. As evangelical Christians they believe in practical outreach. Lionel, who's the communications clerk in Flinders Hospital's emergency department, stood in 2018 as an Australian Conservatives candidate in the South Australian election. Service and values. Which don't stop for weekends or holidays. Janine sees no issue working through. 'It's my service to others I focus on, to free them up to be united with their families. It's a blessing to allow others to spend time away.

'I used to work at the Adelaide Hilton International in my early twenties. A guest asked me if I ever felt like a slave, serving people day in and day out. I can honestly say that I never felt like that. To serve was my joy. I was needed. And my chicken-sitting service has turned out to be something that is needed. And it's a joy.' ●

RELEASING THE DRAGON

INGRID JAUGIETIS

High-school textiles, fashion and food teacher, and Instagram chicken pin-up model

Lives: Canberra, ACT

Pictured with King Henry and Prince Louis, Pekin roosters

'The kids at school always describe me as "the lady with the purple hair". And I say, "Darlings, it's aubergine."'

NUMBER OF CHICKENS:
Fourteen (twelve hens, two roosters)

BREEDS:
Araucana, Barred Plymouth Rock, Dorking, French Wheaten Marans, Frizzle, Pekin, Silkie 'bitsa', Sussex, Wyandotte

NAMES:
Ms Vivian Lay, Ms Zsa Zsa GaBar, Ms Hyacinth Clucket and her sister Sybil Foulty, Ms Blanche Henereaux, Ms Tango, Ms Jazz, Ms Rumba, Holly Go Flightly, 'the two French tarts', top chook Ms Samba and the roosters, King Henry (because he has so many wives) and Prince Louis (son of King Henry, constantly thwarted in his romantic manoeuvrings)

OTHER PETS:
A dachshund, Roscoe, and a rescued sulphur-crested cockatoo, Potato

Of a morning, every morning, before heading off to school, Instagram pin-up model Ingrid Jaugietis takes forty-five minutes to clean out her 'little kingdom of chook pens', as she cannot bear her charismatic flock standing in the overnight accretion of poo. Just on 8 am she releases her romantic leads – King Henry and Prince Louis, two flamboyant yet diminutive Pekin roosters – from their cages inside her Canberra garage. A rooster kept in the dark is a rooster that doesn't crow at first light, or at least is muffled by the walls, and Ingrid is vigilant. She has to be. The substantial fawn outer-suburban house she shares with her husband, fellow teacher Richard, abuts the bush and is backdropped by mountains, but on three sides there are neighbours.

After school Ingrid surveys her bursting closet ('There's chicken maths, and then there's clothing maths and shoe maths') and selects the day's Insta outfit – a vintage playsuit she's sewn herself, a rock'n'roll dress from her clothing sponsor or something va-va-voom ordered online. She chooses her chicken – maybe Ms Hyacinth Clucket, maybe Ms Vivian Lay, but usually one of the roosters as she loves them the most – sets up the camera, and strikes a pose. Then there's footage of the flock to be captured, upskirt shots of the gals to be taken for #fluffybuttfriday, and droll, consciously breathy, *Gone With The Wind*-style copy to be written for both of her Instagram accounts (@poultrystories, @poultrylife). Ingrid is leading her best creative life through the prism of her chooks, and it has been transformative.

'I always wanted chickens, and getting them was my turning point. I loved history and English when I was a kid, and also art and fashion – I studied fashion design. So I'm tapping into what was me from the outset, but it's always been latent. But now, *pewww!* It's like a rocket. For some reason people like to see you holding a chicken. It's a bit left of field, I guess.'

Ingrid began conservatively – four years back, as a newlywed – with a handful of heritage chickens in the rocky, steeply tiered backyard. She gave the chooks playful names ripped from literature and music (Holly Go Flightly is, frankly, inspired), wrote stories and started popping a few photos on Instagram. 'I started with four or five chickens, and that unleashed the dragon.' These days Ingrid has a flock of fourteen, a clothing sponsor, a combined total of 14,000 followers on her Insta accounts, an Etsy shop where people can order her chicken drawings, and business orders for her brooch designs, and was invited to take part in a hot-rod pin-up competition in Ballarat, her first time public speaking. Everywhere she turns is another opportunity, courtesy of her flock. Not bad for a 47-year-old Canberra high school teacher, who has tattoos that she still hasn't quite shown her Latvian mother.

There are boundaries. Ingrid rocks a charming, wholesome, school-teachery vibe. She's in possession of a banging bod, and has statement tatts on her thighs and back (a dachshund, a mer-horse), but Ingrid's not reaching for edge or danger. She has many cardigans. The mini-scenes she writes for @poultrytales are courtly bodice-rippers inspired by the shenanigans in the flock, with King Henry's promiscuous stylings front and centre (followed by second fiddle Prince Louis's attempts to get a look in or a leg over). There's a distinct Anglophile flavour: one foot in Jane Austen, another in Chaucer, and a dash of *Keeping Up Appearances*. Ingrid digs words, history and anthropomorphising chickens. →

A PHOTOGRAPH OF TWO HENS, BUMS UP, HEADS DOWN IN FOOD:

'She swore she would never again take a turn about the room with her French Marans cousin when attempting to woo the resident man in possession of a good fortune.

'She knew it was not advantageous to be so easily compared to the Gallic beauty.

'Her resolve, however, was quickly forgotten when the buffet trolley came out . . .'

A PHOTOGRAPH OF AN OUTRAGED HEN, LOOKING DOWN AT KING HENRY:

'Ms Hyacinth Clucket, being excessively broody and somewhat inconveniently locked out of her nestbox, looked down from her perch.

'"Yes," she mused, her hormones running rampant through her refined being, "He is tolerable, but not handsome enough to tempt me!! I am in no humour at present to give consequence to Kings who are fought over by other women!!"

'Ms Hyacinth Clucket had rather more pressing errands to attend to – namely the "sitting on" of invisible eggs . . .' →

'It's bringing fun, but also the simple life. An innocence, I suppose. Everything these days is so sexualised. I see it a lot in the teenagers at school, how, for want of a better word, they feel that they have to put it out there. Tart themselves up. This is bringing it back down to just being simple again, hamming it up and having a good time. I see these stories play out every day; the pecking order in the flock and the pecking order in the schoolyard mirror each other.'

As you might expect from a craft teacher, Ingrid's philosophical yearning for old-fashioned tropes and mores has a practical underlay. 'I love the 1950s, but I wouldn't have survived it. I like equality. It was a stifling era where a woman couldn't even have her own surname. Where she had to give up work when she got married. No way. I like the romance and the fashion, so I pick and choose what I want.' Richard doesn't share Ingrid's zest for chooks, but that illustrates her point. 'It's my schtick, but he's been fully supportive. He never says no. I think that's what you need most: somebody who lets you shine and helps you along with your own little happiness.'

Ingrid's overarching dream is a hobby farm where she can have roosters without making the neighbours cranky, and collect more, and yet more, beautiful-feathered birds. She's aiming for thirty to forty, but not for Ingrid the lowly ISA brown! She believes in vintage and beauty, that there's a lesson within them for everyone.

'I do like dressing up, but why should we be saving that for a special event? Have a bit of fun every day. Even if you're not feeling the best, when you put on something fabulous it can lift you. The students treat me differently because I put in an effort, and I'm colourful. They'll always comment on my brooches, say they're looking forward to what I'll wear the next day. You don't have to be a cookie cut-out, it's okay to express yourself.'

In the end, Ingrid's a woman for whom chickens answered a question she didn't even realise she was asking.

'I wasn't expecting this. I thought they'd be nice backyard companions, but they're so much more. They bring me back to earthiness, to nature; they ground me when I've had a stressful day. Talking about them connects me to my students; they open up. My son moved out recently, and without the chickens it would be an empty house indeed. And because of the chickens I'm doing so many things now that I would never have fathomed. That's my mantra now: just give everything a go.' ●

HISTORY OF CHICKENS IN AUSTRALIA
A TIMELINE

Australia's chicken history began in January 1788, when the First Fleet landed with eighty-seven chooks. (Today there are eleventy billion[1] of them, so they've done very well for themselves, at least numerically.) An ornament to any kitchen or garden, chooks were found in almost every family from then on. Flocks of a dozen or more were commonplace. They were kept for eggs or – once the fancy took hold in the mid-nineteenth century – to collect and breed for show. They were eaten only once they'd stopped laying, or if there were crossbreed roosters. Chicken dinners were a rare treat.

Then came the tipping point of the late 1950s: experiments with increased efficiency in chicken housing (batteries) dovetailed with the creation of more productive egg and meat birds, and everything escalated. Chickens were hoovered up into commercial industrial facilities – packed in by the hundred thousand – and out of our domestic circle. It became uneconomic and unnecessary to own chickens. Chicken and egg production were industrialised, eggs were now cheap and available all year, and chicken meat plentiful, affordable and convenient. Also, increasingly, frozen and pre-chopped and packaged. Chooks were out of sight and out of mind for a good few decades.

It's good to watch the pendulum begin to swing slightly back, as we belatedly harbour real concerns for chicken welfare. And big agriculture in general. The upsurge in backyard chickens and permaculture has far from peaked.

Turn the page for a handy and faintly depressing historical timeline!

1 Shall we just say 'stacks'? There are approximately 16 million commercial layers and 650 million meat chickens alive in a current twelve-month period. In terms of urban backyards and hobby farms and poultry fanciers' pens and all the rest? Pfft. Stacks. Chicken maths, people.

JANUARY 1788
The First Fleet brought chickens. Obtained at Cape Town. Most didn't make it due to mouldy food and wild weather, but a report sent back to London in May 1788 confirmed eighty-seven chickens. Also eighteen turkeys, twenty-nine geese, thirty-five ducks and 122 wildfowl.

1804
The first reported cockfight and associated melee in Australia. Big night out, mate. It was held at Brickfield Hill in Sydney. Rough part of town, also into dog fights and rat killing.

2 JANUARY 1855
The first verifiable poultry show in Australia, held by the freshly formed Ornithological Society of Tasmania at Pavilion Point, Hobart. A report in the *Colonial Times* newspaper reported that it was well attended, the 'ladies and juveniles much admired' the 'beautiful plumage' of the cinnamon Cochins, and 'there were few instances of inebriety . . . the committee had wisely prohibited gambling tents and such blackguardly pretences for cheating and robbery'.

Nailed it, Ornithological Society!

1857
The first poultry show in Victoria, part of the general show of the Port Phillip Farmers' Society. Or, as it's known now, the Royal Agricultural Society of Victoria.

1850s–1860s
Farmers in the Hunter and Hawkesbury districts put the finishing touches to a locally developed chook, the Colonial Game. Later renamed Australian Game. Sound choice.

1867
The first edition of *The American Standard of Excellence* was published.

1887
The Essendon & Northern District Poultry, Dog, Pigeon and Canary Club was officially formed. Alf Woods's grandfather was an inaugural member.

1894
Lanyon & Sons of Keswick, South Australia, patented their Simplex kerosene-heated incubator.

1905
Chicks were raised in battery brooders in America.

1906–1907
The Hawkesbury Egg Laying Competition was won by 'Chinese' Langshan pullets. Later renamed 'Australian' Langshan. Not sure if this was a sound choice or cultural appropriation.

1922
A team of six Australorps, another Australian-developed chicken, set a world egg-laying record of 1857 eggs over a twelve-month period, or 309.5 eggs per chook. The world went crazy! Orders flooded in! Aussie Aussie Aussie!

1924
World production record was set at Burnley College, Victoria. A Rhode Island Red laid 319 eggs in a year.

1930s
The Great Depression in Australia. Egg production increased as large numbers of the unemployed took up poultry farming.

1932
The introduction of pelleted poultry food.

1940s
The breakdown of chooks used by local egg producers: 66 per cent White Leghorn, 18 per cent Australorp, 5 per cent Langshan, 4 per cent Rhode Island, 6 per cent crossbreeds, 1 per cent other.

1943
The first edition of *Hadlington's Australian Poultry Book* was published by Angus and Robertson. Known as the father of the Australian poultry industry, James Hadlington's classic poultry text had four editions.

1945
The Carter Brothers poultry farm, in Werribee, was recognised as the largest in the world, running a quarter of a million layers.

1948
Complete quarantine ban on importation of all poultry stocks into Australia.

1950s
The breakdown of chooks used by local egg producers: more than 50 per cent crossbreeds.

1950–1951
Industry sources estimate 3 million chickens produced for meat in Australia.

1950s–1960s
Battery cages trending.

1959
Australia's first meat chickens were developed. A Tegel-brand bird.

1960
Broiler chickens marketed at twelve weeks of age.

1968
Australia's first Kentucky Fried Chicken opened in Guildford, Sydney.

LATE 1960s
By the end of the 1960s, annual per capita consumption of chicken had increased to 9.8 kilograms.

1970
Peter Brinkworth of South Australia invented chicken salt. Original recipe included onion powder, garlic powder, celery salt, paprika, chicken bouillon, MSG and Keen's curry powder.

1970–1971
Seventy-five Kentucky Fried Chicken stores opened.

1970s
Breakdown of chooks used by local egg producers: 100 per cent crossbreeds, mostly White Leghorn x Australorp.

By the end of the 1970s, annual per capita chicken consumption almost doubled, reaching 19 kilograms.

1980s
More than 90 per cent of layers now kept indoors, mostly in cages.

1990
By 1990, Australian annual per capita chicken consumption reached 23.7 kilograms.

MID-1990s
The ban on importation of genetic material was lifted to allow the introduction of commercial brown egg-laying hybrids such as the ISA Brown. Game over.

1998
The first commercial-scale free-range chicken brand was launched. It had a 1 per cent share of the market. The first edition of *Australian Poultry Standards* was published.

2004
The red junglefowl's genetic sequence was mapped: the first avian genome.

MID-2000s
Urban flocks started to take off again. Also, hipsters.

2006
Chicken overtook beef as the Australian consumer's favourite meat.

2007
Free-range eggs made up 24.9 per cent of the grocery egg market.

2012
An illegal cockfight was broken up in Rockbank, Melbourne. Eighty-six people were charged, and four hundred roosters 'pumped up on steroids' seized. *Plus ça change*.

2013
The first legal importation of fertile eggs from overseas: twenty new breeds sourced from the UK. Salmonella was discovered during the rearing of the chicks, and the entire consignment was destroyed.

2014
Anti-battery farming television campaign That Ain't No Way to Treat a Lady aired, starring Mick Molloy and rescue hen Little Miss Sunshine.

2016
The second edition of *Australian Poultry Standards* was published.

There were 653 million chickens produced for meat in Australia.

2017
Caged eggs dropped to less than 50 per cent of the market share for the first time.

2018
Free-range chicken made up 20 per cent of the market.

2019
Publication of *This Chicken Life*!

LUCKY

AMANDA CLEMENTS

Paediatric nurse, mother of twins, warrior

Lives: Hobart, Tasmania

Madeleine and Amanda pictured with Lucky, a Barnevelder rooster

'When things change you get the chance to dictate the terms you live the rest of your life by.'

NUMBER OF CHICKENS:
Six (five hens and a rooster), plus a few chicks

BREEDS:
Araucana, Barnevelder, Rhode Island Red and mixed heritage

NAMES:
Lucky, Doctor Evil, Agnes, Edith, Psycho and Little Brown Hen

OTHER PETS:
A black cat, Ninja

'The chickens are great pets because they have zero expectations of Madeleine. They like to have a chat.'

Tempting though it is for a well-meaning human, there are compelling reasons not to help a struggling chick hatch from its shell. 1) The effort it takes to free itself of said shell is an integral part of a chook's prep for a healthy life; if you assist, it will be weak and (probably) die. 2) It (very) possibly has developmental issues and should (and will) die. 3) It may not have finished absorbing the yolk into its body, and will die. 4) If you pick shell away from the chick, or try to crack it out in any way, you will rupture the blood vessels lining the shell, which are still attached to the chick, and the chick will bleed out and die. The other life-altering consequence is that if it survives after your intervention, damned if you're not going to fall in love with it.

Amanda Clements, paediatric nurse, ex-occupational therapist, mother of twins, smart cookie and experienced small-time chicken hatcher, knows all this. She does not assist with a hatch. Uh-uh. Particularly not a small incubator hatch, where things go wrong all the time. 'I prefer using a broody hen. It doesn't often go very well with an incubator. You have problems.'

Two years back, when the broody hen wasn't broody, Amanda used an incubator and one of the chicks was stuck in its shell. 'He was very large. Too much chicken and not enough room. He was shrink-wrapped.' Chicks use their 'egg tooth' to peck a hole into the air pocket at the end of the egg, and then, shifting in a circle, they 'unzip' the top of the egg and tie off the blood vessels. 'He couldn't get around. And I was like, okay.' In she stepped. Amanda mimes her next move, holding the egg up close, so fragile, and giving it her full, rapt attention. 'So I got a toothpick, and held the egg, and just very gently did the unzipping round the edge of the shell.' *Pock pock pock*, she goes, recreating every tap with the point of the toothpick. 'I perforated the rim, sprayed it with water to keep him moist, and he did the rest.'

There was a reason he couldn't get out. The little fella's feet and legs were, in Amanda's words, 'all stuffed'. The legs were badly splayed, the feet curled up, not right at all. 'I did what every good and responsible person does and went on YouTube. And I thought, shit, that's easy, I do this for work. So I made him little bell-shaped splints out of wood, and I had Coban adhesive wrap, and I splinted his legs. He looked at me like, "I hope this is okay. I'm trusting you."' His hips were sorted after three days, his feet – mmm, not so much. But he could walk, and Amanda named him Lucky. 'Beautiful Lucky. My lovely boy.' And lucky this imperfect Barnevelder roo is. He's a treasured pet now, ruler of the roost; the rest of the hatch, all male, fit and healthy, ended up in the freezer of this self-sustaining home. 'I have a weakness for damaged animals now.'

Amanda Clements is softening. She can't be as armoured as she once was, when she worked as an occupational therapist in New York post 9/11, with the city she loved gone and her friends traumatised. 'I was pretty hard for a long time – that's why I came home to Hobart – but my children have softened me. Having a child like Madeleine has softened me immensely. I'm much more forgiving of human frailty. We all stuff up. No one's perfect. If I hadn't had a child with a disability I would have been far more "practical" about that chick. That's how I would have put it. I would have said, "Yeah, right, but this is not viable," and I would have culled it.'

•••

Madeleine, Amanda's ten-year-old daughter, fraternal twin to lively brother Sam, has a full-wattage smile, a gentle nature, a tight posse of schoolfriends, and lots of cheek. When she met her heroes – the Wiggles – backstage at their Hobart concert in 2018, she placed her hand directly on Yellow Wiggle Emma Watkins's heart – 'She did that to all of them, to connect' – and then noticed Emma's sunshine-bright bow and absolutely went in for the steal. Towards the end of our interview, she drifted over from the TV, where she'd been laughing at Pee-wee Herman. Madeleine glanced at me, then sweetly, meaningfully, waved 'bye-bye' a few times. Amanda hooted. 'Just because she's got a disability doesn't mean she's not rude.' Madeleine loves Pee-wee Herman, she loves the Wiggles, and she loves the chickens. The five-acre garden is being worked on to ensure that, despite leg splints and worsening mobility, Madeleine can still get easy access to the chooks in the back. →

'The chickens are great pets because they have zero expectations of Madeleine. They like to have a chat. If she has food, that's awesome. When she was younger she talked more, and she'd be standing in the middle of them, chatting away, and they'd just surround her. When our new chicks are ready, they'll go outside her bedroom window. She'll spend a lot of time there, at the window, watching and talking to them.' The chickens are helping them both: good company, good eggs, good memories.

It was only in 2017, during some clever surgery to address foot spasming on Madeleine's weak left side, that Amanda received a definitive genetic diagnosis of Madi's condition. 'I got this phone call. Again, moments in your life. So the nurse said to me, "Oh, I'm going to get the senior geneticist to talk to you about this. It's good to have a chat with someone with experience." And I've worked in health care long enough to know. I thought, "Shit, it's something serious."'

Madeleine did not, as they thought, have cerebral palsy but a rare (500-cases worldwide rare) spontaneous chromosomal disorder called BPAN, where iron builds up in her brain. It is untreatable. 'It's degenerative. The iron builds up and up and up. Once puberty hits, and Madeleine's started early, they develop Parkinsons-like symptoms.' Absolute best case scenario, lovely, cheeky Madeleine makes it to her mid-forties.

Amanda and Darren, her then partner, had tried, all stops out, for a child. After their eighth failed IVF transfer, she remembers feeling flat, and saying to him, '"It doesn't matter what kind of child we get. I don't care if they're disabled or not. I just really want to have children." I don't know if you tempt fate. It's just that the universe said, "Well, okay, cool, we've got one for you."' So many moments where reality shifts on its axis. Sam and Madeleine, twins, were born after Amanda's thirteenth IVF treatment. One of them was the IVF implant, and one conceived naturally. What are the odds? There's no distinguishing which child came from which egg.

In 2015 Amanda was nursing her father, who was dying of bone cancer. In the space of two weeks Darren left, her father died, and she lost her hospital contract. In the darkest of times, the chickens have seen her through. 'When Dad was dying, my house wasn't a farm, but there were chickens, and moments of reprieve to get into the garden and chat to the girls. Get some sunshine. After he died I'd have days where I was up to my eyeballs in study and misery, and the chickens, they were just all over the yard. They're just – they're just very good companions. The way they look at you, you know, with their heads to one side. I love that.'

Amanda's grandmother, Mona, had chooks. And her dad was a farmer. A man of his time and place who farmed up Tasmania's north-west coast. 'He came to visit me, walked up the path and up to my cat, and started talking to him right away. He didn't know I was there. "Well, aren't you a handsome fellow," he was saying, and I thought, "Well, that's where I get it from." I've always talked to other people's chickens. That whimsical love of animals is something I get from my father. He bred cattle, and was a serious man, but he'd get teary over animals and I do too.'

The days are good at the moment. Demanding, but good. Sam's doing well, Madeleine's still Madeleine, and the chooks are moulting but enjoying their new role as permaculture heroes in Amanda's expanded vegetable garden. Lucky is shagging the daylights out of Little Brown Hen and continuing to provide quality companionship.

'He should have been in the freezer, a big chap like that. My handsome boy with very bad feet. Ah, but there are my feelings about the worth of any being who is different. I have an absolutely fascinating child, and she can be challenging, but she is a beautiful soul. There's a purity there you don't find every day. She brings a lot of joy into the room. And Lucky, he was just this glorious-looking boy and I couldn't do it. There was just something very special about him, the same as there is with Madeleine. He's perfect to me.'

Amanda glances towards the window. The latest clutch are on the patio in a coop, getting a dose of Hobart cold to acclimatise them. Toughen them up. They're only bubs, but Amanda's already fretting that there are a couple of roosters. 'There's an interesting shift happening. Being connected to Lucky, now I'm looking at the ones I *think* are roosters, and I don't know if I can cull anymore. Once you accept their individuality it's too hard. I've got a feeling the kids and I are going to be vegetarian. Yeah. Once you change, there's no walking it back.' ●

THE HENSIONERS

MARIE MORRIS, NAOMI SMITH AND PAM WOOD

Living in aged care, chicken lovers

Live: Whiddon Group aged-care facility at Bathurst, New South Wales

Pam pictured with Princess and Henrietta, Silkies

'Oh, I never watch television. I watch the chickens.' Marie

'There shouldn't be favourites. It's not nice.' Pam

NUMBER OF CHICKENS: Four (hens)

BREEDS: Silkie and Silkie cross

NAMES: Princess, Henrietta, Pride and Olaf (named after the snowman in *Frozen*)

Pam Wood, seventy-nine, lies in bed, propped with plumped pillows, a small bright pool of a woman in a sea of bedclothes, her room a homey explosion of teddy bears, ornaments and orchids. She's not well at the moment. In pain. Leanne Bowland, the facility staff member assigned as chicken champion, wheels in the pink trolley containing Princess and Henrietta. 'Hello, my babies,' Pam says, stretching out her arms. 'Come here, my babies. How are you, my friends?' Leanne pops the Silkies down on the bed, and Pam rests her hands in their fur, one on each chook. Princess and Henrietta relax into, and under, Pam's hands. Together, they all drift off to sleep.

...

Every morning Marie Morris, seventy-nine, makes her way to the same place. A chair next to one of the floor-to-ceiling windows at Kelso aged care, which commands a view over the coop and gardens. Everyone knows it's Marie's chair. She's appointed herself guardian of the chickens, and it's a job she takes seriously. She doesn't take her eyes off them, watching them fossick in the undergrowth, lay eggs under bushes, quarrel and dirt-bath, and if the staff don't lock them up at 4.30 pm (as happened a memorable once), she refuses to go down for her tea until they do.

...

Sipping a latte, Naomi Smith, eighty-nine, sits in Kelso's outdoor cafe area, enjoying the warm shade with some other residents. The chickens, interested to see if there might be a bit of banana cake heading their way, move between her legs and brush up against her as they cluck around. She laughs and notes how lovely it is, just having chooks about. The others agree. She recounts how her great-grandson dived under the fence and found two eggs when he visited. 'He didn't break them. And he was only three!' Someone else wonders where Pride's got to. Half an hour later, they're still chatting about the chickens. →

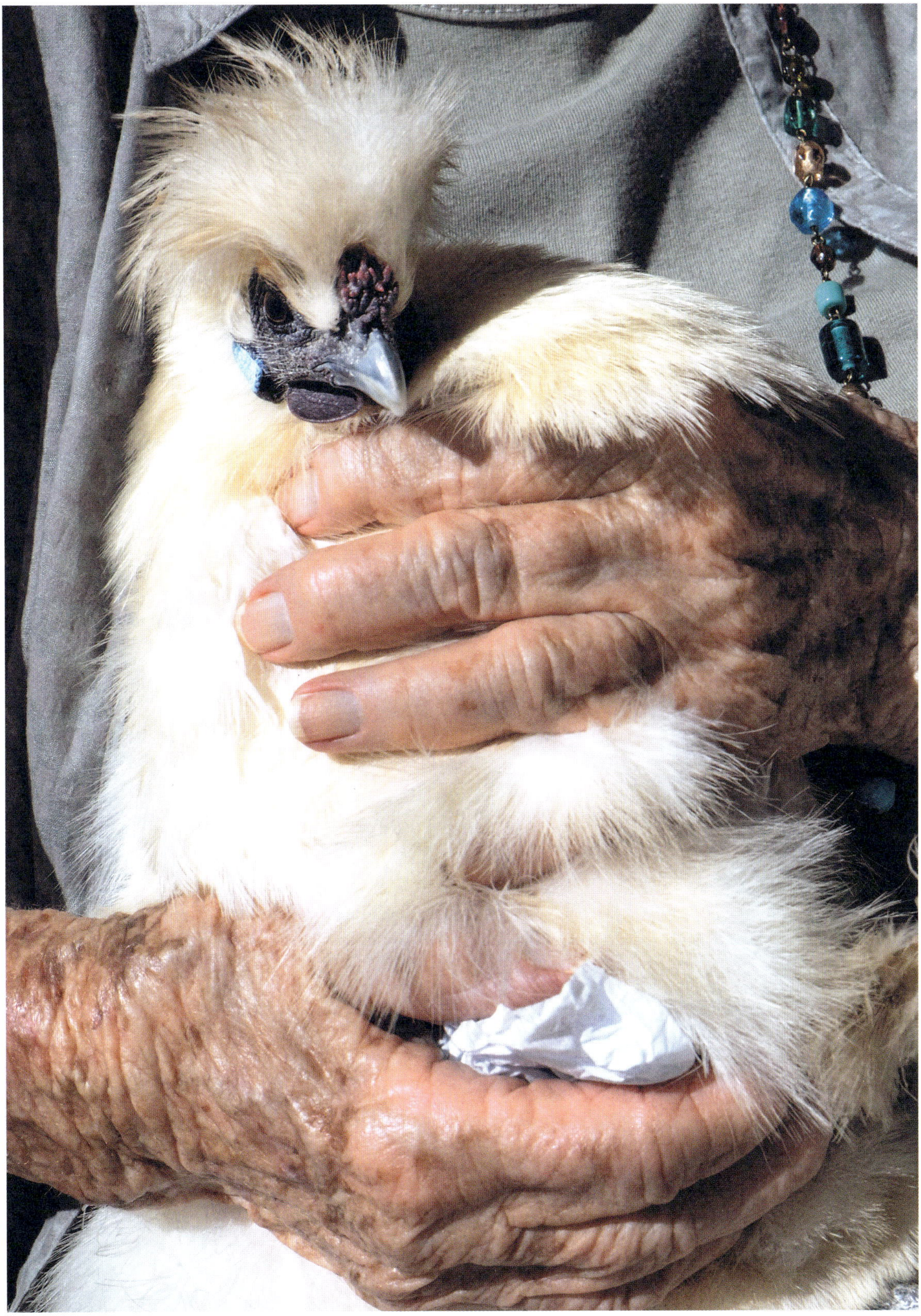

'It gives people something to come out of their rooms for, and they're much happier than the ones who don't.'

The Whiddon Group's HenPower program is evident throughout the fastidiously clean, cathedral-ceilinged, sun-bright facility in Kelso. There's a photo of the 'resident of the month' holding Princess. Chicken-shaped blackboards display comfortably hoary chicken jokes ('I just ordered a chicken and an egg off the internet. To see which one comes first! . . . I'll keep you posted.'). There's chicken art, certificates won for chicken art (third prize at the Royal Bathurst Show for *Gran's Button Jar Rooster*), chicken craft, chicken pillows, a plush chicken at reception, chickens wrought from bright enamelled iron. But mostly, of course, it's the chickens themselves, either outside scratching around or inside sitting in their trolley, magnetising residents and visitors alike. Princess, Henrietta, Pride and Olaf may be small fluffy chooks who are mainly concerned with food and flock politics, but they're punching well above their (bantam) weight in terms of contributing to the residents' wellbeing. They have transitioned Kelso from a facility – albeit a very *nice* facility – to a home.

Naomi, dignified and switched-on, was already in situ when the chickens were introduced to Kelso and notes that the facility has changed. 'They're just . . . lovely. It's made people come closer together, given them a bond, I think. A shared interest for the community. Something to talk about. There really wasn't that interaction before; we kept to ourselves more. We mix now.'

Like many of Kelso's residents, Naomi grew up in the area, and her parents, with eight children, had a flock of about twenty White Leghorns. Bog standard for Naomi's generation. Bringing chooks into the facility has meant a reminder of life outside aged care, and a rekindling of memories, as well as the opportunity to create new ones.

'You've given up everything to come in here. You're away from everything you've known,' she says. 'Having the chickens, it reminds you of your home, it's comforting. It just sort of makes things more natural, more homely. And my great-grandchildren love them, they're excited to visit.'

'Excited to visit' are not words you usually associate with small children and aged care. But the chooks, they light up everyone.

The HenPower program came via Equal Arts in the UK, a creative ageing charity based in Newcastle with a charter to improve older people's lives through creativity. HenPower was designed to combat loneliness, and 'hengage' (their phrase) aged-care residents with arts activities and chicken-keeping. A more-bang-for-your-buck combination of pet therapy and creative ageing. Five years ago Karn Nelson, the impressively titled Executive General Manager of Strategy and Innovation at Whiddon, heard Equal Arts present at a conference in England. She was impressed and intrigued. 'A lot of our facilities are in regional, rural or remote areas, and people who grew up like that had chickens and really value access to animals. I got permission to trial it.' HenPower was adapted, assessed, and is now in half of all Whiddon's facilities. The results have been beyond what Karn hoped for.

'To be honest, I'd looked at it more as a form of creative expression and arts therapy, something for people to get involved in. That's what it is in the UK. I was very surprised at how much more it brings than that. It's having something to care for, and bringing normality into their lives. It builds social connection. Gives them purpose. The stroking is calming and soothing. And it's so *easy* to have chickens. So cheap. We measured the quality of life benefits, the build in self-esteem. We measured depression through the Cornell scale, and the rates went down dramatically.'

Marie grew up with chooks and ducks in Wallerawang, and her daughter, Donna, works for WIRES, New South Wales's largest wildlife rescue organisation. 'We've all been animal mad,' Marie admits. She also acknowledges that she 'feels like a chicken mummy'. When the chickens are in sight, Marie can't tear her eyes away. She's obsessed, and it would be a cruelty, actually, if the chooks *weren't* there. One of Marie's big treats is watching the weekend staff trying to herd the chooks to bed at 4.30 pm. They haven't learned they need to lure the chickens in as it's before dusk; they chase them every time. 'It's a three-ring circus!' Marie chortles. 'You can't tell them. I say to Leanne, "They need the red feed bucket, then they'll come," but no, you can't tell the men. They're chasing them around! Oh, we laugh our heads off.'

HenPower differs from other successful programs, such as dance or art therapy, in that it doesn't have to be timetabled or programmed. The chickens are just there, twenty-four seven, part of the woof and weft of the facility. They're reducing agitation in residents with dementia, causing incidental exercise because people walk outside to see them, creating friendships, bonding the staff further to their clients, luring the frail and aged from their rooms. Leanne says the chickens know when someone's ill – 'They know what their job is' – and settle in to provide comfort. Pam is clear on that too: 'They cheer me up. They cheer us all up.'

Once in a while, a chicken dies – there was once a Prejudice, for example, sister to Pride – and in an aged-care facility, where death's whisper is close, the chooks normalise that conversation, the reality of the cycle of life. Ultimately, though, what the chickens bring to Kelso are vitality and life.

Naomi is a social being. Happy to cuddle and stroke one of the chickens, she also values engagement with people. She notes this: 'It gives people something to come out of their rooms for, and they're much happier than the ones who don't.'

The HenPower trial won a best practice award for Whiddon: vindication given Karn's fear that they'd 'get into trouble' for integrating chickens. 'Imagine! In such a highly regulated environment! We thought we'd be told they were dirty, that they're a risk. But the quality agency who does our compliance heaped praise on the program. Because it's transformational. It wasn't difficult at all, in the end, because the best aged care isn't about creating a "hospital environment"; it's not just about "clinical needs". Or antiseptic. It's about what's good for the residents. It's important to break the norms, otherwise you live in fear of sanctions. A home with chickens, well, it feels vibrant.' ●

MOTHER-MORPHOSIS

LIZ SKITCH

Clown, theatre devisor, director, sometime nude accordionist, reluctant chicken convert

Lives: Preston, Victoria

Pictured with Rachel/Astrid, an ISA Brown

NUMBER OF CHICKENS:
Four (hens)

BREEDS:
'Speckledy ones', ISA Browns

NAMES:
Trevor and Helen, Rachel and Mike. 'They're named after our favourite street performers, because they're pretty free range.'

OTHER PETS:
A Siamese fighting fish, named The Fish, as in, 'Have you fed the fish?'

Liz Skitch, host of the Moulin Beige comedy cabaret, co-artistic director of deBASE production company, children's performer, and custodian of an Imeldian collection of mumus, has always had chickens. Scratching around in the backyard of share houses in Queensland, and later Melbourne. Now they occupy a large run near the bees in the rambling permaculture garden she shares with her husband, trumpeter Donnie Stewart, and their two small children, Molly and Fonzie. Despite Liz's theatrical bent she'd never call the chickens feather babies. No way. They're just chooks, bought from the market: the critters that supply the eggs, great for the kids. 'I always thought of them as an arm's-length animal. A bit smelly. Not attached at all,' she says.

Then she began work on a solo show, *Mothermorphosis*, a Kafka-inspired, comedic performance-art piece that deconstructed motherhood. Liz and her director, Maude Davey, decided the show definitely needed a chicken. Okay, then. Liz surveyed the flock, set up an obstacle course in the garden, and posted footage on Facebook of the chooks auditioning for the role of 'Astrid the chicken'. Friends were invited to vote. The personalisation process began. 'Everyone thought Helen would win because she's the prettiest, but she didn't have the performance chops that Rachel did.'

It took a while for Liz to warm to 'Astrid'. She realised she was a bit scared of chook handling, and kept putting off incorporating Astrid into the show. 'It's like when you have a role where you're performing naked, and you keep putting off the moment in rehearsal when you actually take off your clothes.' But every day she was packing Rachel/Astrid into the cat carrier and putting her in the car, and every day Rachel/Astrid was in the rehearsal room, and something shifted.

'I started talking to her in the car, which wasn't intentional: chatting away. And then I realised I needed to give her treats, to work out ways she would be a happy chook. That I *needed* her to be happy. In *Mothermorphosis* I was the mother, and she was symbolic of the child. And this transition happened. When I held her, I realised she picked up on my energy. If I relaxed, so did she. It's the same energy transfer as when you're nursing a child. I started feeling protective. MY chicken. She literally opened the show, and was onstage with me the whole time, and it took the pressure off. I saw her differently. By the end of the run I was taking her out afterwards into the foyer because I wanted her to meet people! She was pretty much taking a bow with me.'

Liz was surprised by her attachment. She still hasn't told Donnie. 'He's old-school. He'd laugh at me.' Rachel's back in the chook run with the others, but she's 'Astrid' to Liz. Something has permanently changed. 'As a child I enjoyed tending the graves of our pets in our garden. My budgie Amigo had an elaborate grave, to the point where the neighbourhood kids thought it was a person. The cross just got bigger and bigger. With Astrid, I'm definitely going to be sad when she dies, and I wasn't expecting to feel that. She'll have a very well-tended grave.' ●

RED PILL

WHY I NO LONGER EAT CHICKEN

I can't tell you the last time I ate chicken. A couple of years back, I reckon. The last time for anything is the same, if you've not actively made a stand. At some point you look in the rear-vision mirror of life and go, hey, where did doing the splits go? I used to love doing the splits! (I could never do the splits.) Or picking a scab? (Still do this, soz. I know: disgusting.) Or pashing a random dude at a party? (It's been eons! Those were the days!) If you knew it was the final time, you'd have a chance to mark it. You know, reflect on your choices, write an affirmation, light a frigging candle. But mostly we don't, and the caravan moves on.

There are pros and cons to making a big deal of a decision. It can be a line in the sand but, as all New Year resolvers know, it's often an invitation to fail. The only thing that feels worse than addiction to cigarettes, for example, is the ciggie you light less than a day after your theatrical public announcement that you will never, ever, no way, absolutely *not* smoke again. I gave up the smokes midway through the 1990s, and I remember why and how. I remember waking at 7 am with the dizzying revelation that I could just give up, and how exultant I was. But I have no recall of the final fag. There was nothing special or climactic about it. No 'this is it' moment accompanied by a tearful letting go. Given where I was in my smoking career, it would have been a reflex bedtime rollie held between nicotine-stained fingers, and I would have been hating myself. Fair dinkum, who wants to remember that?

Same same with chicken. Gwyneth Paltrow talks of 'conscious uncoupling', and good for her but there's a lot, I reckon, to be said for unconscious uncoupling. The stuff that happens without us particularly thinking about it, behind the scenes. A whimper rather than a bang. I have a suspicion that my final chunk of chicken was probably finger food. An opening night probably (I direct cabaret shows), perhaps a party. No dinner on my part, and a bit of panic that I was drinking unpedigreed red wine on an empty stomach which leads, as sure as eggs are eggs, to migraine. So, yep, thanks, I'll take that skewer. Pretty much the same way everyone who eats chicken eats chicken, then. Grabbing it unthinkingly because it is there. Convenient.

If it *were* that particular skewer, I do recall that it tasted of sadness. Chicken is uncontroversially delicious – the albatross which chooks carry round their necks is that they taste of magic and everything wants to eat them – and I was as surprised as anyone to have an 'ashes in my mouth' experience. What was my problem? It was free range, I always checked for that. I did a pivot years back, after reading Jonathan Safran Foer's *Eating Animals*. A quote on the cover from *Time Out* stated 'Read this book. It will change you', and – fair warning – it did, as Safran Foer is a sneaky genius when it comes to sucker-punching the reader about industrial animal farming in America. I emerged from its pages a somewhat smug 'ethical omnivore'. Which is to say, only eating meat if I knew its provenance and that the animal had led a free-range, organic, happy-go-lucky life attended by hand-maidens and had owned two investment properties.

Good for me. I live in inner-city Melbourne; this is not a sacrifice. It is an excuse to visit more farmers' markets and hipster cafes. I'm a one-woman episode of *Portlandia*. The one who points at the menu, raises her eyebrows and enquires pointedly, 'Is this free-range?' (You would be amazed, by the way, how often the answer to that question is, 'No, but it's gluten-free.' An equivalent question-answer pairing would be 'Do you sell scooters here?', 'No, but our cars are painted red.')

Being an ethical omnivore has been a total 'get out of jail free' card. If I paid the dosh and bought free-range and organic chicken, then I could proceed free of guilt and responsibility. When I got my own gals I deftly handled the cognitive dissonance of loving them, admiring their smarts and personalities, and sporadically eating their relatives. I'd probably still be straddling the 'occasional Sunday roast' fence right now, if I hadn't written this book. Research leads to knowledge, and once I understood how we've modified broilers – our generic 'meat' chicken – I couldn't walk it back. A broiler could be living large in the actual Garden of Eden and still have a short, brutish life. Well, that's just *great*. That sound you're hearing is my illusions shattering. →

The science magazine *Cosmos* published an article in December 2018 outlining how the modern chicken, specifically the broiler, is the ultimate representation of the Anthropocene period of earth's history. The Anthropocene is right now, the era defined by human intervention and impact, manifesting in such delights as climate change and species extinctions. Researchers from the United Kingdom have pointed to 1950 as the start of the Anthropocene, citing as a key marker the chicken-breeding competition launched by an American supermarket chain to create a fatter chook. The post-war competition was called Chicken of Tomorrow, and the end result is our broiler, a bird engineered to be slaughtered between five and seven weeks of age. It is constantly hungry, rapidly puts on leg and breast muscle, which leads to a decrease in the size of other organs (particularly heart and lungs) and a skeleton which collapses under its Schwarzeneggian body weight. If they're not slaughtered as babies (not even two months old – jeez), they invariably die early of cardiac issues or organ collapse. This is every commercial meat chicken, so free range,

organic, schmorganic. Doesn't matter. They're Frankenchooks, hatched to live in pain then die. Billions of them, every year, worldwide. So, I . . . yeah. Stick a fork in me, I'm done. (The irony, according to friends who rescue, is that broilers are the friendliest of chooks, and really *like* people.)

There's very little that's comfortable about peering behind the curtain at any industrial animal-production process. Layer chooks (your ISAs and Hy-Lines), have been similarly modified to deliver an egg a day until they're eighteen months old and then be dispatched. That's the business model. Laying every day puts a huge strain on their bodies and, unsurprisingly, they have short lives plagued with reproductive issues. Still: eggs, amirite?

I'm comfortable using my showbird Rosecomb and Silkie eggs. They're not frequent layers, and the eggs are small, but with eight chooks it's plenty. I'm trending vegetarian, though – veganesque even – because now I know too much. No big statements, because damn, I love cheese, but I may well be looking in my rear-view mirror one day and thinking, 'Huh, where did eating animal products go?' ●

LOVE BIRDS

SEBASTIAN GALLOWAY AND EFFIE ROSE PRYER

Fine artists, bird lovers, quiet types living the dream

Live: Fern Tree, Tasmania

Sebastian pictured with Buster, a Sebright rooster; Effie pictured with Muriel, a Black Modern Game

'You can tell the ones Sebastian has named. They're ridiculous. Scone had a son he called Sputnikolas.'

NUMBER OF CHICKENS:
Twenty-ish, a few young'uns (mostly hens, four roosters)

BREEDS:
Araucana, Barnevelder, Black Modern Game, Cochin, Gold Laced Sebright bantam, Silkie and Polish

NAMES:
Muriel, Buster, Dave, Parsnip, André, Bentley, Scone, Evil Scone and Yeti, Anna and Trousers, Zelda, Lucy and Hyacinth

OTHER PETS:
Three gobblers (male turkeys), Leonidas King of Sparta, Jemeremy and John; three female turkeys, Tabitha, Sybil and Sybil; a guineafowl, Josh; a rescue Abyssinian cat, Puss; and a shelpie (Shetland sheep dog), Jeremy

Seb looks at Effie. Effie looks at Seb. Seb speaks first. 'Yeah. We're not massively extroverted. Probably massively the other way, to be honest with you.' He holds Buster, a golden Sebright rooster, loosely in his lap. Buster's paramour, Muriel, a tiny, sweet, inky Modern Game bird, sits absorbing light on Seb's shoulder, for all the world part of the painting behind her. A black-ops shadow chicken. 'I embrace it rather than try to fight it. I like spending time in my study, just working on my own things. And with the animals, the chickens, you do feel you're in the company of others. It definitely doesn't feel lonely up here, that's for sure.'

There's a pause. Muriel moves her head. Buster doesn't. Effie, Seb's girlfriend, also a visual artist, picks up the introversion baton. 'It's hard to socialise sometimes when you think, "Oh, I have to drive all the way down the hill."' Seb nods. She continues. 'I find myself a little bit isolated, but it's so lovely here. A little haven. I don't want to live in the city. I can go hang out with the chooks, which is still being part of a social network. They're just not the same species.' There's another pause. Effie laughs at herself. 'Wow. I sound really sad. I need to get out more.' →

Sebastian Galloway and Effie Rose Pryer are two artists – oil painters: Galloway specialising in birds, Pryer in nudes – nestled in an old-school rental in the wilderness suburb of Fern Tree, 420 metres above sea level, and two thirds of the way to the cloud-wreathed peak of Kunanyi/Mount Wellington. Professionals in their field, they both work at Mona, Effie as its conservation supervisor, Seb in technical operations, building exhibition spaces and installing large-scale works. Mona, the Museum of Old and New Art, employer of about five hundred driven creatives, is the engine room of Hobart's reinvention as a throbbing centre of cultural cool. It could be read as ironic that Seb and Effie, Taswegians born and bred, often prefer being in the company of their chooks and menagerie of fauna to, say, socialising and networking now that Tasmania's capital is finally going off like a cracker in a nutsack. 'There's a sense of pride here that wasn't here before,' acknowledges Seb.

But they value aesthetics and solitude. And they're lovebirds, and they love birds. In their warm, light-flooded artists' house, wooden display cabinets hold found treasures: a bird's nest complete with teeny egg, an animal skull, an antler, a pearl rosary with a crucifix, wooden bobbins coiled with coloured twine, piles of eggs, a bowl of rocks, all just so. Curated and conserved. Effie has arranged bouquets of brilliantly coloured macaw feathers in aged glass bottles. One of Seb's paintings, a small portrait of Nugget, one of the first chickens they 'liberated' from Mona's resident flock of chooks, leans on the wall next to the television.

'We both really love birds,' says Seb, whose body of work includes minutely detailed paintings of Tasmanian wedge-tailed eagles, scarlet robins, goshawks and fairy-wrens. 'We love animals, but particularly birds. They're a passion for both of us. And I guess chickens are a really accessible option for somebody who likes birds and wants to experience their company. You know, interact with them.'

There's been some jostling over who gets to paint which chicken, which pet. Seb shotgunned Puss, their ridiculously glamorous rescue Abyssinian. 'I'm not allowed to paint her now,' Effie says. Effie got in first with Muriel. But then Seb used Muriel, and her boyfriend Buster, as models for serious works. Both moody, gorgeous portraits sold, for $3000 each, although not to the same buyer. Seb's slightly sad about that, clearly fond of the paintings.

'They looked good as a duo. It's definitely something to capture, the personality of these little chickens. You know that one particular look that they make, where you know what they're thinking? Muriel's got this really good little curious face, when she's discovered something new, and that's what I wanted in that picture. Buster's a really tiny chicken, but he's got this huge fierce attitude. He thinks he's about twenty times bigger than he actually is. I wanted him to come across as a chook with attitude.'

It's probably fair that Seb made the big splash with Muriel and Buster, as they were a Valentine's gift to him from Effie. 'It was a little bit bold considering we'd been together less than six months at the time,' says Effie. 'I just went, Seb likes chickens and we don't have any of those.'

They'd been on a date to the Royal Hobart Show, and seen a cage of Modern Game birds. Seb was captivated. 'I was fascinated by them. The most ridiculous-looking birds I'd ever seen. These long-legged, slim, tiny chickens!' Four years later, Buster and Muriel, Seb and Effie are all still together.

As is the way with chickens, the numbers multiplied. The house came with enough rooms for Seb and Effie to have a painting studio each, as well as chicken sheds and runs, and an abandoned silver rooster they named Bentley (the prior tenant bred Araucanas). At Mona, it was easy to get permission to appropriate a bunch of chicks from the museum's large free-ranging flock. Nothing fancy. 'Just chooks,' says Seb, who was probably responsible for naming one of them Nugget. From there they got guineafowl, and two turkeys, 'which multiply by themselves, as it turns out'. They acquired Araucana hens for Bentley's benefit, some Silkies – 'Dave's amazing. Like a little white angel,' says Effie of their Silkie rooster – and then staged another chicken heist at Mona. 'I kept seeing this sort of silvery-grey girl,' says Seb. →

'Oh, she was nice. So one night as we were leaving work we saw two of them sleeping on this perch, and we were like, "Let's get them! I'm having this chicken!" We drove all the way home with these two little chooks on Effie's lap.' They took the friend as well, and named them Beautiful Chook and Friend. As these things go, the chicken they ended up loving was Friend. 'Beautiful Chook was a bit of dud, to be honest.'

For all they bind Effie and Seb to home ('If we're away we don't really think of coming back to our house,' says Seb. 'It's like, "Okay, let's go back to the chooks."'), having chickens has, sneakily, made them more outgoing. Given them an outreach topic of conversation. 'Yeah. It's extroversion amongst introverts,' says Seb. At Mona Seb has gradually become known as the Chicken Whisperer. The Chook Guy. 'I get this all the time: "Oh, you'll probably know – I've got a chook question."'

Effie concurs. 'They're always asking him questions. And he doesn't hold back when people ask him how the chickens are going. Gives them the full rundown. Gets the phone out. He loves it.'

But, then, there's always something to talk about. So much striking animal eye candy, so much drama. Beautiful Chook died of a snake bite. Friend in a mysterious perching accident. Most of their Silkies were taken by a quoll, which picked them off the perch like low-hanging plums. 'They roost low,' mourns Effie, who then shyly recalls Seb chasing the quoll around the run in the middle of the night in his underwear. Muriel, Buster and the kids have to live under netting due to the unwavering attention of local goshawks, eagles and owls. 'Raptors everywhere,' says Seb. There are rabbits in their garden, grown sleek and fat on stolen premium chook pellets. A pademelon. A wombat. The turkeys, who are flat-out hilarious; the three males sailing around like feathery galleons, gobbling in outraged unison, snoods wobbling.

'You can't look out of a single window without some kind of animal drama going on, to be honest,' says Seb. 'It's constant entertainment.'

Why on *earth* would you go down the hill? ●

THE STANDARDS

ALF WOODS

Icon, empire builder, in thrall to the fancy

Lives: Greenvale, Victoria

Pictured with a bantam Indian Game hen

'I talk chooks all the time. You never know who you're going to meet. I've got some good memorabilia that way.'

NUMBER OF CHICKENS:
Two hundred

BREEDS:
Old English Game, Indian Game, Modern Game, Pekin, Buff Cochin and Sumatra

NAMES:
Ha ha ha! No.

OTHER PETS:
His granddaughter's dog – 'She calls it a crossbreed, I call it a mongrel. But she reckons it's a beautiful dog, and it is.'

'I just loved chooks. Wanted to be mixed up with poultry. That's been my whole life. It still is.'

Alf Woods, ninety-two, reckons he's got another ten years in him. I reckon he has too. Vigorous and trim, flicking a comb out of his trouser pocket and through his dapper hair before being photographed, he presents as twenty years younger. A charmer. He can also crouch and chase chickens like a champion – handy when you've got hundreds of chickens, mostly hard-feather game birds, zipping about. He lives on five acres in Greenvale, with 132 exhibition pens, his daughter, a granddaughter and two grandsons. Moved there from Essendon in 1970 after one of his birds flew a fence and pecked holes in the neighbour's tomatoes. He works five and a half days a week at his business, Blue Sea Pools and Spas, pulled back from seven after he turned eighty. He just loves work. He reads voraciously, mostly old books on poultry. Doesn't rate the new ones. 'How to build a coop is not information.' Eats chicken soup most nights. Of course he does. Ex-boxer and ex-swimming teacher. Of course he is. There's discipline here, sharp steel. It's why Alf still rules the roost in the Australian exhibition poultry world.

He's a judge and exhibitor, sure, with a river of ribbons hanging from his office ceiling. But he's also President of the Essendon Poultry Club, President and life member of the Game Club of Victoria, a director of the Victorian Poultry Fanciers Association – Australia's peak body – and is on the Australian Standards Committee, the group that produces the *Australian Poultry Standards* (APS) for breeders and judges. A lot of important meetings take place around Alf's kitchen table.

Known as 'The Standards', the APS is a 350-page hardcover tome that contains each recognised breed's standard of perfection. It's the bible for the poultry fancy in Australia, and without it, there is no exhibiting. Alf's still playing the big game, still got the vision; his lifetime goal is to secure the financial independence of the Victorian Poultry Fancy. He wants to ensure it continues into the future. Alf understands legacy. And he has concerns. Once he's gone, who will maintain the standards?

'If you don't have the standard, for any kind of an animal, where would it finish? You wouldn't have horse shows, you wouldn't have dog shows, you wouldn't have poultry shows. If you don't have standards, well, you'll finish up with nothing. If you haven't got a process in place to maintain something, it just deteriorates. We have a standard in our own life, too. We must keep to it. Try and improve things for everybody else.'

Alf has love of poultry in his blood. His father, Allan Woods, was an exhibition poultry man in Essendon, and his grandfather, Alfred Woods, was an exhibition poultry man in Essendon before him. Young Alf didn't stand a chance. He won his first trophy in 1940, a cup engraved 'Best Standard Novice Ancona'. Still has it. He has a room chock-full of history and memorabilia. 'I think we follow in the footsteps of others, and I follow in the footsteps of my family. When I was little my father would take me to bird nights at the Essendon Poultry Club. I probably slept most of the time, but I insisted he take me. My uncle had a milk round in Ascot Vale, and he'd drive me on his horse and cart through the gate into the Royal Melbourne Show. I'd be straight off the cart and into the poultry pavilion. I just loved chooks. Wanted to be mixed up with poultry. That's been my whole life. It still is.'

Alf's a poultry man to his bones. A slave to the fancy. He married his love, the red-headed Gladys, on 1 April 1950, and on 2 April they departed for their honeymoon. At Sydney Airport, Alf mentioned there was a slight change to their plans, hailed a cab and swept his new bride to the poultry pavilion at the Sydney Royal Easter Show, where he spent the first day of his honeymoon with the chickens and his chook mates. Alf has no clue what Gladys got up to. 'She didn't love fowls,' he notes. A benign acceptance that Alf will always prioritise poultry seems to be the norm. His daughter Pauline, who lives next door on the Greenvale property, wanders through the kitchen. 'We're just talking chickens,' he announces. Pauline doesn't respond, but the look on her face says, 'Tell me something I don't know.'

Talking chickens is the foundation of Alf's life. He's well loved and well connected, and has a lifetime of friendships to show for his passion. 'You have to share your interests. If you haven't got friends, you've got nothing in life. That's what I value most, the friends.' Alf's travelled the world, meeting other chook fanciers. Japan, America, Saudi Arabia, the UK. Wherever he goes, he stays with chicken people. 'They're always good, kind people.' In Japan he sought out the Onagadori, a rare and decorative historic chicken which has a tail up to ten metres long. He took photographs of a Japanese collector's taxidermied birds, unwinding and fanning out their tails. 'He said to me, "You made them come alive. Because of your love of animals, you'll live a long and happy life." It made me emotional. I was recognised by him. I cried. And I never cry.'

Alf was formed in a different era. Harder times with stricter rules. One of his warm recollections is of a stern lecture from a breeder, because he hadn't removed the previous year's scales from the legs of his Old English Game cock. '"Alf," he said. "When you visit friends with your family, do you wash behind your ears, your face and hands? Polish your shoes and put your best clothes on? Then never let me see you exhibit a fowl that you have not bothered to properly prepare." I was fourteen years old. It was a great wake-up call.'

Alf's drive and self-reliance make sense when framed by his upbringing. No softness allowed. Working for his father's plumbing company (apprenticed at the age of fifteen), he fell ten metres from the Melbourne Showgrounds roof when he was twenty-five, did a complete somersault, and landed on his back. 'I was so lucky, I should've died. In them days, you know, you didn't call an ambulance. They put me in the back of the plumber's truck and carried me into the clinic.' The doctor told him to go home and take some Aspros. He was in bed for ten days. Then he caught the bus and walked three hundred metres to the shop. He asked his dad for some money: the mortgage was due, the baby needed feeding. His father said, 'If you can walk from Essendon Station to the shop, you should be back at work. When you've done five days' work, young man, you'll get five days' pay." I was back the next day. He was hard.' →

Another parenting tip from Allan Woods. If you find out your twelve-year-old son's been smoking, take him down the back of the house, get a slat from a wooden banana box and belt him. 'He said, "My boy, you'll never smoke again." I didn't. That's why I'm still alive. Today, he'd go to jail. But he'd given me, I believe, another ten or twenty years of life.'

Alf is no-nonsense and all rigour, which is old-fashioned now. And there's change in the poultry scene. Chicken people aren't necessarily coming through the ranks of the fancy. There are people with 'feather babies', people who post with their chickens on Instagram wearing outfits, people with hybrids, people with crossbreeds. Alf knows they're there. 'Ask the feed merchants. We've got one in our club. They're selling two pallets of poultry feed a week to backyarders.' But an appreciation of chooks is the only intersection in this particular Venn diagram. It's a different world, and one where there's a collision of sensibilities. Alf was invited on ABC radio as an expert for a discussion on poultry. He sat and listened, and the calls were coming in. 'I colour my Silkies with a hair colour and they're pink.' Another rings up. 'My name's Mandy and I've got a blue one!' Alf's mind boggled. Then it was his turn. 'In history, every backyard in Australia had a dozen chooks in it, and they ate the scraps that now go in the dustbin. And my father, like most people, when it came to Christmas dinner, he'd go and pick out the oldest rooster he had, then Mum would pluck it, it'd be baked, put on the table, and it was that tough we'd need a chainsaw to cut it.' Thus ended Alf's radio career. 'Thank you, Mr Woods, that's enough.'

I wish I'd heard that broadcast. Their loss. It's never enough. Alf can talk under wet concrete about chickens. He knows everything. 'I could chatter on forever about chooks. Love 'em. Pauline came with me to judge a chook show in Dunedin in New Zealand – she insists on accompanying me now – and went off to town while I finished judging and made a few comments after the show. She came back, and one of the ladies asked her, "You're Australian. Is that your dad in there, the judge?" I'd been talking for an hour and a quarter about life in the chook world, and hadn't said one thing about their poultry.'

Alf always talks chooks because 'You never know who you're going to meet.' It's his way of ferreting out other chicken people, forming those connections. Bonding. Most of the time he can tie them back into his network. They'll be the daughter of someone he knows, a cousin. He collects a lot of memorabilia that way, too. People are glad to pass it on to someone who cares.

Alf keeps saying, 'I'm lucky. So lucky.' It feels that way. He misses Gladys (who died after an issue with her pacemaker in 2015), and weeps when he talks about her, but his family live with him. Another part of his legacy. 'I take my little grandson, from the age of two, to the national park out the back and we look at the birds and we look at the grass. I'm trying to educate him the same way as my family educated me. To a love of nature. I hope I succeed. I've got six great-grandchildren.' He pauses, wistful. 'Maybe one of them will follow in the footsteps of Alf.' ●

'My father, like most people, when it came to Christmas dinner, he'd go and pick out the oldest rooster he had, then Mum would pluck it, it'd be baked, put on the table, and it was that tough we'd need a chainsaw to cut it.'

GIVING IT HER BEST SHOT

TAMARA KENNEALLY

Photographer, chicken rescuer, storyteller, activist

Lives: outer Melbourne, Victoria

Pictured with Neptune, a Leghorn

'When chickens are treated poorly they're like shells of nothing. They're all the same, blank stares. They don't know who they are. But take them out, treat them well, give them a month and they become who they were always meant to be. They become themselves. It's beautiful to watch. It's one of my favourite, favourite things.'

NUMBER OF CHICKENS:
Eleven (hens)

BREED:
Mostly ISA Browns, all rescues

NAMES:
Bobby Bob Bob (named for zookeeper Bob Fossil from *The Mighty Boosh* because her head bobbed up and down), Bowie, Babushka, Kek Kek, Goblin, Neptune, Irma (aka Wonky Wobbles II), Strawberry, Hyacinth Bucket, Red Robin and Cruskit

CHICKENS FOSTERED OUT:
Popeye, Almond, Rhonda, Speckles, Arrow, Pisces, Aries, Rapunzel, Judge Judy, Lavender, Leo, Marigold, Moonlight, Moonshadow, Nanatoo, Nipper, Skipper, Zippy, Thumbelina, Oda Mae Brown, Crazy in the Coconut, Walnut, Storm, Star, Rosemary and Miss McCraw. 'They all get names. All of them.'

OTHER PETS:
One cat, Whiteface; two koolie dogs, Molly and Jackson; and seventeen sheep, Right, Junior, Pancho, Cosmo, Limpy, Angry, Alice, Heidi, Screamer, Moss, Elsie, Frodo, Banjo, Scrunky, Trouble, Greedy and Wrinkles

Tamara Kenneally, photographer and animal lover, rescued her favourite girl, Bobby Bob Bob, from a caged egg farm in March 2014. 'She's special. When we're going through the cages we have to choose, and we choose the worst. Her neck was very bad. We took her out and she looked up at me and made a hopeful noise, and that was the start of me and Bob. I went, okay, okay, this is my chicken.'

Bob's got broken vertebrae in her neck, a screw in her gizzard and one drooped eye, and her lower beak is cut in half. Bob also had a dire respiratory infection which left her with a wheeze like a pug's, and she was bald, her feathers pecked out by her frantic cage-mates. She has, like most ISA Browns, reproductive issues. The wars? Bob's been through 'em. Tell you what, though, she may not be picture perfect, and vets said she wouldn't last a year, but she's a curious and happy chook. Her hobbies include screaming for attention, eating mealworms, sleeping inside, modelling for social media photos, and helping Tamara do Important Work on the computer. Still with us (an impressive six years old), loving life and, unsurprisingly, Tamara. All of them do, of course. Tamara's boutique, special-needs rescue flock is billeted in a chicken wonderland in her parents' back garden in the outer suburbs of Melbourne, and she moves among the chooks with gentle hands and gentle voice, a beacon of kindness and respect. A Florence Nightingale of chickens.

Tamara, forty, is true to her nature. She was that gal at school who crazy-loved animals. 'Yeah, I used to ride around the streets on my bike looking for lost dogs – they were never lost, they were just wandering outside their houses. I had a little "save the animals" kit on my bike and off I'd go. Mum and Dad would be, "Oh no . . . is she gonna bring home another cat that just lives down the street?"' The animals Tamara rescues today genuinely need help, which is probably why James and Verlie can't put their foot down. Tamara indicates a tiny, defiant circle of fenced-off green. 'Dad's been trying to save some grass. No chance, really. When I go out at night with this rescue friend, if I'm here, Mum'll go, "Don't bring any more chickens home!" and I'm like, "Oh, no, I won't, Mum." And then a week later she's like, "Are there chickens in the courtyard, Tam? No!"

'I feel like I'm the polar opposite of the people who breed them to show. For them it's the most beautiful, the most perfect chicken. For me, it's the discards. The more broken they are, the more disabled, the more likely I am to love them. I'll go anywhere. I'll get a call – "I've got a chicken. One leg, one eye, no feathers," and I'll be, "I'll take her! Where are you?", and they'll say, "Darwin!", and I'm like, "I'm on my way!" I'm an absolute nutbag.'

The majority of Tamara's rescue animals, mostly sheep and chooks, live at Lefty's Place, an iconic animal sanctuary in regional Victoria which she founded in 2010. Named after her (now deceased) pet lamb Lefty, it's the public face of Tam's consciousness- and fund-raising. There are only eleven chickens at her parents' place, the most stricken, the ones Tamara saves in hope of bringing them to health, and counter-balancing their experience of torture with kindness and freedom.

'These are the ones who need medication daily. Most of them have had broken legs due to rough handling, and I don't want them jumped on by a rooster. Neptune came from an abattoir, and both her legs and wings were broken.' Neptune now perches happily on Tamara's shoulder, but all the stories are beyond grim. Babushka had her eye pecked out in a cage. Bowie was skin and bone after having been abandoned in a free-range egg farm depopulation. Cruskit had broken legs and was bald and emaciated. For chooks softly lifted, broken-bodied, from hell, this place is a miracle. This is what Tamara is in it for.

'There's something incredibly rewarding in giving them a little bit of time, just a little bit. It's that they love life so much, I think that's what it is. They just get so much out of life every day.

'I feel most connected to these girls, I think, because they're not quite right and because they try so hard. They try so hard to walk, and sometimes I feel like I try so hard to walk too. It's a heartbreaking journey, honestly, because they die so young, and they die constantly, even if you give them all sorts of vet care. There are a lot of little graves here, a lot of plaques.' →

Tamara doesn't like to call herself an activist. It's a contested and competitive term in the vegan and rescue community. Yet she is. Rescues are part of what she does, but her vocation is raising consciousness around the costs to animals, the suffering caused by commercial farming. 'I want people to know they do have personalities, they're not just egg machines.' She helped obtain the footage used by ABC's *7.30* investigation into the Star Poultry abattoir in 2017. Four months of filming showed chickens' feet being torn off, chickens being scalded in boiling water while they were still alive. Some of Tamara's more contentious photographs are from free-range chicken farms, which aren't necessarily the chicken fun palaces we hope they are. Tamara gives a voice to animals, and witnesses the unseeable.

Her photographic portfolio divides neatly into two categories. The yin and the yang. The light and the (very) dark. There are the 'after' shots, taken at Lefty's Place, of animals embracing their life of freedom after being rescued from death. Chickens jump for joy, goats snuggle, cows graze. They're idealised, heartwarming images which celebrate each individual critter, and are for sale as cards, calendars and prints to raise funds. Food and treat bills are, as you'd expect, never-ending. Tamara's vet bills are catastrophic.

Then there are the 'before' shots, the Hieronymus Bosch horror show which is Kenneally's activism. Terrified cows with bursting udders waiting for death at abattoirs, sheep crushed into trucks, and chickens. So many chickens. Caged chickens, broken chickens, bloody chickens, insane chickens, chickens drowning in faeces and mud, chickens left to starve. Kenneally documents and portrays the life stories of the byproducts of commercial farming. The images are bleak, rending and accusatory, not what anyone wants to consider as they tuck into a poached egg or parmigiana. Which is, of course, the point. It's hard to face the suffering of animals, and no one knows that more than soft-hearted Tamara, who looks at a great deal of it.

'You can't save them all. And you can't take them all to the vet to be put down. There's thousands in these farms you have to walk past who are suffering, and you know they're going to die. There was a really bad cage farm once – every single cage had girls who were done. They were on their way. What do you do? My friend said, "Should we break their necks right here, right now?" I said, "I can't do that, I can't possibly do that." And she said, "I can't either, so what do we do?" So we leave them.'

Encouragingly for Tamara, there's definitely an upswing in demand for rescue birds. People are increasingly interested and aware, and keen to adopt healthy rescue chooks when they become available. 'We can't keep up with demand. People are essentially good, I think. They want to help.' Most of Tamara's work, though, is with the chickens that are rescued purely to be euthanised. Or with her own small, compromised flock. She makes hard decisions all the time, and has learned not to go too far in trying to save a life. There's no point in forcing a big operation on a damaged chicken, not when quality of life is your aim. It's a devastating vocation she's got for herself.

'Yeah, sometimes I do sit there and I cry, and I think, "What am I doing? I could be travelling, drinking cocktails every day, making lots of money." But I'm here with these chickens trying to patch them up to live a little longer. And to keep them alive, it's like sticky-taping a bloody glass bowl together. You're sticky-taping constantly, constantly. And then one day they just go nah, their body's done. And it's the saddest thing.'

Tamara describes herself as a crazy chicken lady, but she's not crazy. She's just doing her best. It's tougher since her partner, Justin, died suddenly last year. An ex-rocker, his body just gave out, too. Her comfort, as always, comes from within the rescue community.

'We all support each other, and when one of our girls dies it's always the same things we say to each other. "At least she got to live." That's what we say. She got to live, and she got to die peacefully. She didn't die in the slaughterhouse. Not having her throat slit or going into the boiling water alive. She got to die loved. She was buried. We all say that to each other, it's all the same. And we all cry, yeah. But chicken people, I think, are beautiful people.' ●

'There's something incredibly rewarding in giving them a little bit of time, just a little bit. It's that they love life so much.'

EGGSEGESIS

AN INFORMATIVE, HAND-PICKED, NON-DEFINITIVE GLOSSARY

ALBUMEN: The white of the egg. Contains about forty different proteins, and is the bit that turns into the chick if the egg is fertilised. Yeah, the yolk's the power supply.

***AUSTRALASIAN POULTRY*:** Megg and Jessamy Miller's definitive go-to magazine for all things fowl. It is scandalously affordable, and contains no-nonsense, informative articles about poultry in very small type. It is pleasurably like going back in time.

***AUSTRALIAN POULTRY STANDARDS*:** The official breed standards for poultry fancy in Australia. Contains the standards of perfection for breeders and judges alike. If you're considering dipping your toe into the fancy, the standards will guide you.

BANTAMS: Small chickens. The word really does refer to size. From many a conversation I've come to understand that most people think bantams are Silkies. Silkies *are* bantams, it is true, but not all bantams are Silkies.

There are two types of bantam. 'True' bantams, which don't have a full-size counterpart. So, yes, Silkies and also Rosecombs, Sebrights, Pekins and Belgian d'Uccles. Then there are 'miniature' bantams, which are pocket versions of regular chickens, such as a bantam Rhode Island Red, bantam Game or a bantam Australorp. They are all adorable.

BLACK GOLD: Chicken poo. It's that good for the garden.

BLOOD SPOT (on yolk): This categorically does not mean that the egg is fertilised. It is not chicken embryo. You can eat it. Sometimes there's a rupture in a blood vessel during the formation of the egg. Totally normal.

BLOOM: The protective coat on a freshly laid egg. Don't wash it off! The egg has up to 17,000 pores, and the bloom seals them. It helps keep the egg fresh and bacteria from getting in, and reduces moisture loss from the egg. If there's dirt on the egg, just brush it off.

BLUE: Grey. Why are grey animals called blue? It's a mystery. Anyone with an answer, I'm keen to know.

BRAINS: Chickens are a whole lot smarter than we give them credit for. The 'chickens are stupid' trope only began once they disappeared into battery farms, and we stopped interacting with them. How convenient. The smart chooks are the fun ones, curious and naughty, getting into scrapes like the heroine of an Enid Blyton novel. The ones with personality and charisma. Interestingly, of course, we don't breed for that. We only breed for 'meat production' or 'eggs' or 'beautiful plumage'. What if we bred chickens for intelligence? We'd be in trouble. These are the thoughts that keep me up at night. →

BROODINESS: When a chicken wants to hatch eggs, a fog of hormones descends and she goes broody. It can be a bit tricky to distinguish 'sick' from 'broody' if you're a new player. When a chook goes broody, she stops laying eggs. A broody chicken is almost certainly on the nesting box, maybe on eggs, maybe not, flattened into a feathery pancake, in a trance with a thousand-yard stare. This is why you can put hats on them. They have no idea. They can maintain this for the twenty-one days it takes to hatch an egg. They will growl if you try to remove them, and also make adorable *bok bok bok* noises. If you're keen to hatch chicks, get yourself a Silkie, a Frizzle, a Cochin, a Sussex or even an Old English Game. They love to be mamas.

BUMBLEFOOT: Not, as it sounds, a minor, ineffectual wizard in *Harry Potter*. It's an infection caused by a particular type of bacteria getting into a small wound on the bottom of the foot. More common among heavier birds. Very painful. You can go online and watch a video about how to dig out the infected plug and drain the pus. Or: that's what vets are for.

CHALAZAE: Not, as it sounds, an Instagram-famous rapper. You know those ropey strands of egg white that make you go *bleah*? Yes, those are the chalazae. They're part of the egg architecture and twist in opposite directions to anchor the yolk in place. They're normal - the fresher the egg the more visible they are - and are most definitely not developing embryos.

CHICKEN HANDLING: Handle your chooks regularly. Once a week at least. The best time is either before you let them out of their coop or once they've gone to bed. There is nothing more fruitless than chasing your chickens around. They hate it; you look like an idiot.

We do ours once a week in the morning, before we let them out of the coop. We've got scaly leg mite in our run, and it's a challenge to totally eradicate it, particularly with wild birds flying in. What you can do is dunk your girls' legs regularly in oil (it suffocates the mites), and at the same time give them a once-over. Have they lost weight? Are they healthy? How are their eyes? Their combs? Are their happy little vents winking at you? We consider it 'cuddle time', and there's no doubt that the more they're handled the happier they are to be handled.

CHOOK: Australian vernacular for 'chicken'. It's presumed to have developed from the English practice of saying 'chuck chuck chuck' to call chickens, or the Irish practice of saying 'tioc, tioc, tioc', to call chickens ('tioc' means 'come' in Irish).

CHOOKAS: Australian theatre speak for 'Good luck!' What you say to a performer on opening night, or before going on stage for any gig. It began back when chicken was an expensive meat, and unaffordable for most actors. The stage manager would scan the audience and, if it was a full house, report back that 'chook was on the menu'.

CLOACA: What a great word for the tremendous and versatile vestibule into which the intestinal, urinary and genital tracts open. AKA the 'vent'. AKA the chicken's one-stop shop orifice for everything. I suspect it's the etymology for 'clacker'.

CLOACAL KISS: How chickens 'do it'. When a daddy rooster loves a mummy chicken very much, etc.

CLUCK NORRIS: Exceptional name for a rooster.

CluckAR: Augmented Reality Free Range Egg Detector - an app from CHOICE, so it's backed with actual research. It scans the barcode on a box of eggs and tells you how 'free-range' those free-range eggs actually are, and whether or not you should buy them. Comes up with a little animation of chickens to illustrate the stocking density. Very cute, very useful. Kids LOVE it.

COCK: Short for 'cockerel', a term these days used in the show community to refer to a young rooster. 'Cock' originally came from the French *coq*, but to be honest the American puritans found it all a bit embarrassing and switched the language to rooster. Which also saw 'weathervane' replace 'weathercock'. Humans are ridiculous sometimes.

COMB: The red (usually) thing on top of your chicken's head. Depends on the chicken, though - Silkies have black combs. The most common type of comb is the 'single' comb, and that's usually what we draw if asked to draw a chicken, but there are eight shapes altogether, depending on the breed. There are 'pea', 'buttercup', 'strawberry', 'rose', 'V', 'walnut' and 'cushion'. You can tell when a hen is coming into lay by the colour of her comb. If it's bright and healthy, she's laying. If it's a bit lacklustre, she's moulting, not well or just off the lay.

DAME EGGNA: Excellent chicken name.

DEBEAKING: Ugh. A beak is not just a chook's mouth, it is also its hands and weapon. Debeaking is a practice used in industrial farms to stop traumatised chooks from attacking each other. It's flat-out cruel.

DIRT BATH: Pig-Pen from *Peanuts* was onto something. Fluffing their feathers in a pile of dry dirt is how chickens bathe. It gives them great pleasure and keeps mites and parasites to a minimum. A basic need for chooks. Provide a dust bath in an old tyre or similar, or if they free range they'll find their own spot.

DOGS: Do you have a dog? Excellent. I've heard many stories from people whose dog died, or the neighbour's dog moved away, and next minute the fox got the chickens. The smell of dog deters foxes. It's not a guarantee, but it raises the bar. There are particular breeds of dog that will protect chickens, such as the Maremma - that's what it lives for and how fabulous. But most of us, statistically, are rocking a something-doodle named Sebastian or a rescue staffy cross with the spatial awareness of a tank, and they will do nicely. Just make sure they're introduced to your chooks slowly, and are trained to understand that they are *not* to chase them or eat them. This is key. An untrained dog can lose its mind and go through your flock in a flap-inducing frenzy and savage the lot. Once they've worked out that this is NOT an option, it's all good. You've seen plenty of YouTube clips of unlikely animal friendships, right? Domesticated animals can get on just fine if the boundaries are laid down early and firmly. That said, if you've got a dog with a strong prey drive, fences are also good.

DUMPING: Here's the thing. Chickens are domesticated prey animals; they rely on us for food and safety. It's not okay to 'return them to the wild' or deal with your unwanted rooster by dropping him off at the local park or tip. Particularly in Australia, where they're not indigenous. It's cruel. They will be taken by dogs or foxes, or lead frightened, hungry lives. If you're hatching chicks, you need a plan for the boys, which means either: knowing you can keep any roosters yourself (hello, overnight box in the garage or laundry), having somewhere specific lined up to take them, paying the vet to euthanise them (costs about $30 per bird), giving them to someone willing to cull them or culling them yourself. Eating your own rooster is far kinder than any chook meat you can buy, so stop making that face (vegans, at ease: obviously not talking to you here). Essentially, while many of us enjoy our chickens as pets, they are also livestock, and come with a side order of pulling up your grown-up farmer pants and taking responsibility.

EDWARD LEAR (1812-1888): English literary nonsense poet who popularised the limerick. Best known for penning 'The Owl and the Pussycat'. Another of his indelible poems was 'There Was an Old Man with a Beard':

There was an old man with a beard,
Who said, 'It is just as I feared! -
Two Owls and a Hen, Four Larks and a Wren,
Have all built their nests in my beard.'

Reminds me of Costa.

EGG COLOUR: Depending on the breed of chicken, eggs can range in colour from white or cream through to blue, green, pink, nut brown and speckled. A lot of folk buy Araucanas simply because they lay pale blue eggs. The colour has no bearing on the taste of the egg, just on how pretty they are.

EGG TOOTH: A tiny horn-like protuberance on the upper beak of a chick. It's used to break the chick out of the shell, and then falls off. The egg-tooth fairy then puts a dollar under the chick's pillow.

ENRICHMENT: Enrichment is the new buzzword in backyard chook husbandry. There's a definite movement towards adding mirrors, swings and xylophones to the coop, or creative ways of offering treats, such as sweet corn on a rope, or bread and veg hanging from a laundry drying rack for socks. The idea is to keep your chickens stimulated, healthy and happy. If your chooks are regularly free-ranging, it's not particularly necessary, as they will keep themselves perfectly occupied, thanks, but if you have to keep them cooped, well, how would you like a bare room for hours on end? →

You'd probably quite like Netflix or a book (ha ha), some music or a gym. Chooks are the same. They have mental health needs, and if it's just them and nothing else, you'll end up with a bored, cranky flock with behavioural issues. Give them a cabbage on a rope and some interesting perches, and they'll be entertained for hours.

THE FANCY: A collective name for fancy fowls and their enthusiasts.

FEET: Chicken's feet are surprisingly soft underneath and sensitive. The scales are on the top.

#FLUFFYBUTTFRIDAY: An Instagram hashtag. Every Friday, chook lovers worldwide post photographs of their chickens' feathery behinds. It's a stunning display of fluffy chicken-butt pulchritude. May also contain traces of fluffy cat butt.

FREE-RANGE EGGS: The Australian standard for free-range eggs is a stocking density of a maximum of 10,000 hens per hectare, with meaningful and regular access to an outdoor range. Producers must also prominently display the stocking density on the packaging. Figures published by the CSIRO and backed by the RSPCA say that a stocking density of no more than 1500 per hectare would be more in line with what people *think* free-range means. You know, happy hens frolicking in the grass. Choose accordingly.

FREELOADER: Affectionate term for an adult lady chicken who is eating all of the food and laying none of the eggs. Often seen looking unrepentant on social media accompanied by many 'crying with laughter' emojis. The ultimate complaint is, 'I had to go the store!' Common in the colder months, or when she's moulting. Completely normal. Give her protein and let her enjoy her cloacal holiday.

FRIZZLE: A self-explanatory feather mutation. Adorable. There is a recognised Frizzle breed, but the Frizzle gene is also present in other breeds. Hence a Frizzle Polish, Frizzle Cochin or Frizzle Pekin.

***GALLUS GALLUS DOMESTICUS*:** Latin for chook. The red junglefowl from the Burmese jungle, from which all chickens are in some way descended, is *Gallus gallus*.

HORMONES: There's a persistent belief that chickens are fed hormones in industrial farms. Not so. The practice was given a shot back in the 1950s, but it turned out to not fatten chickens, and was phased out in the 1960s. Along with lava lamps and the Watusi.

HUGS: A chicken hug is when your chicken winds her (or his) neck around yours.

LANGUAGE: Chooks have somewhere between twenty-four and thirty unique vocalisations. They are a very chatty species, with different sounds for 'hello', 'back away from the food, Shirley', 'that's a freaking hawk – get undercover' and (the loudest) 'check out the egg I just laid, it is amazeballs'.

LAYER: Technically, a hen that is currently laying eggs. A 'good layer' is one who lays a lot of eggs.

LAYING EGGS: Chickens slow down their egg production in winter as the days are shorter. Chickens need twelve hours of daylight a day to lay, because their reproductive cycle is triggered by the natural light cycle. It's possible to manipulate them into laying more frequently during the off season by using artificial light, but it's unnatural and another way of wearing out your chickens. It can also mean that if you get point-of-lay pullets (see next page) late in the season, they may not start laying until spring!

LONELINESS: Chickens get lonely. They're a social animal, and they're not designed to be by themselves. They may also be lonely in a flock if they don't have a buddy.

LONGEVITY: How long do chickens live? Depends on the chook. Commercial hybrid layers such as ISA and Hy-Line essentially lay themselves to death. Because they've been manipulated to produce so many eggs, they often succumb to reproductive illnesses such as egg-yolk peritonitis and reproductive tumours. There are cases of old lady ISAs, and that's fantastic, but four is usually the upper age limit.

Heritage breeds, which are 'natural' layers and have been bred and raised naturally, can expect to live to about eight. But I've known folk with chooks who are in their teens. The basic life-extenders are: good health care, good nutrition, good housing, a good vet and maybe some cuddles. In the past, when chooks really were left to their own devices, not much effort went into extending their lives. Now that it's more of a priority, they'll indubitably be living longer.

MOULTING: You wander out to your chicken coop of a morning, and it looks as though a pillowcase has exploded in the run. One of your chooks has no tail. Another is basically nude. Has a fox attacked? Are they diseased? No, it's moulting season! It comes on in cooler months, which seems counterintuitive. Some chickens moult hard and look like someone's glued feathers from a mangy boa to a wilted balloon. Others, like my Silkies, moult like a snow dome, but with no discernible difference to the eye. Completely normal. Will come, probably, with an easing off of egg-laying, because they're using all their protein to make new feathers. A good time to give them extra protein.

NESTING BOXES: The recommended ratio of nesting boxes is one per four or five chickens, but no matter how many you have, no matter how generous your chicken-to-box ratio, the chickens will only want to lay eggs in one. They will queue. Fight. Protest. Lay an egg on top of another chicken. Lay side by side in an adorable mop of feathers.

There are three schools of thought. One: chickens are conservative. We're talking about their future children here. It's like private school. If one chicken's picked a box, if it's the *traditional* box, then it's safe, tested and proven. They all want in.

Theory number two: They're trying to collectively get a clutch of eggs happening, because the more quickly they aggregate a hatchable number, the sooner a broody hen can sit on them, and the sooner there are chicks.

Theory number three: They like laying together. Give them a bigger box.

NUTRITION: You'd think you could just throw chooks any old thing, what with all the 'they eat your scraps and turn it into eggs and black gold' rhetoric. But the fact is that humans have played merry heck with a few breeds, and the humblest-*looking* of the backyard chickens, the ISA or any other laying hybrid, requires more protein than fancier breeds due to the pimping of their reproductive systems. →

It takes a lot out of a chicken to lay an egg a day, it's not natural to lay through the moult, and if you don't give them layer pellets they're probably not getting what they need. I know. It's very frustrating. But they've been genetically modified to need a boosted level of feed. Scraps are good too, for tastiness and variety, but don't think you can skimp. Modern chooks are not what they were. They're next gen. I've chosen to go grain, myself, but I don't have hybrids.

OFF THE LAY: Chickens tend to go off the lay, or slow down, in winter. They need a certain number of daylight hours to trigger laying.

ORNAMENTAL BREEDS: Chickens kept because they're decorative, not because they're good for eggs or meat. They still lay eggs, of course, they just haven't been maximised for it. Most bantams are ornamental, and any breed with extravagant feathers. The ones that make you go 'ooh'.

PENIS: Apologies in advance, NSFW, but I'm overly fascinated with the fact that roosters don't have them. *I know.* Blew my mind too. The only birds that do, actually, are ducks, swans and geese. A drake penis is epic, corkscrewed, barbed, explosive and between 15 and 40 centimetres long, and in all seriousness is why you shouldn't house ducks and chickens together. Damage can absolutely be done if a horny drake is indiscriminate in his affections.

Roosters, the ancient and international symbol of fertility, have, like the ladies, a cloaca or vent. Congress is achieved through a 'cloacal kiss', where the rooster's and hen's cloacas form a seal, not unlike a space-station docking, and the rooster deposits what's known as his 'packet' of sperm. Fascinatingly, if the hen is not into that particular rooster, eyeing off a more dominant roo, she has the ability to eject the packet up to twenty-four hours after mating. She sort of spits it out of her cloaca. By the by, if you Google 'chicken pretends to be dead to avoid rooster', you'll find a compelling video of a chook *playing dead* to avoid a rooster's romantic attention. We've all been there. Henimism, amirite?

The lack of a penis is considered an evolutionary ploy by the rooster to make sex more acceptable to the female. (See, as a counterpoint, 'explosive barbed duck peen' above. Female ducks have evolved corkscrew vaginas with multiple dead ends to avoid unwanted fertilisation. They tense up their muscles to skew the entry of the unwelcome penis. Drakes grow a new penis each season, and the more competition there is, the longer it gets. Yeesh.) Instead of competing in a sexual arms race, roosters have gone the non-penetrative, sensitive new-age route. In vitro, a penis does begin to form on a male chicken embryo, but at the nine-day mark a chemical is released which shrivels it away.

Okay, I'm done now.

PIN FEATHERS: After moulting, the feathers start to grow back. And they're strange, spiny things that make it look as though your girl's being hen-pecked. No, it's just the spines of the feather coming through first.

POINT-OF-LAY (POL): A girl chicken who is on the cusp of life's great mystery. You know, (whispers) *about to get her period*. She'll be eighteen to twenty-four weeks old, the age at which most of us buy new hens. Buying POL chickens doesn't guarantee they will lay immediately. Every chicken is different, and they often need time to settle into their new homes. Unhappy, stressed chickens don't lay. Unless they're hybrid layers, which have no choice.

PREDATORS: The chickens' greatest vulnerability is that *everybody* finds them delicious. In Australia the main predators are foxes. But, in no particular order, they are also often taken by dogs, hawks, eagles, pythons, dingos and quolls.

PRINCESS LAYER: Another excellent name for a chook.

PULLET: A young hen, less than a year old. From Le French *poulette*, meaning 'young chicken'.

ROO: A colloquial term for rooster. Originating in America, it has caught on among the backyard chook community here. It's a tad confusing in Australia, given we call kangaroos 'roos', and it's certainly not used in the show community. But it does fit in with Straya's determination to shorten or infantilise any word, by force if necessary (see 'bikie', 'firey', 'pollie', 'ambo' and 'hostie'). It's also affectionate, which fits in with our rekindling love of the show-pony male *Gallus gallus*.

ROOSTER: Originally short for 'roosting bird'. Obvious when you think about it.

SADDLE FEATHERS: Long pointy feathers that grow on a rooster between his tail and wings at about three months old. At the risk of being Captain Obvious, they grow in the shape of a saddle. One of the easiest ways to spot that you've hatched a roo.

SCALY LEG MITE: My nemesis. One of the many mites that chickens can cop. Often flown in by wild birds, it's a weeny mite that burrows under the scales of the legs, feeds and leaves droppings. It makes the legs look scabby and crusty – because they are – and is painful as well as unsightly. Gah. There are many types of mites, another reason to handle your chickens regularly so you can spot them.

SICK CHOOK: Hey, so, chickens do pop their clogs quite regularly. They can live long and happy lives but equally can collapse mysteriously and suddenly in a foof of feathers – here today and gone the next. They're hardy *and* vulnerable: highly resourceful and adaptable, able to scratch by on a minimum of food and resources if necessary, yet when ailing go downhill fast. They're a prey animal, which means they're highly skilled at cloaking illness. Consider them the opposite of children who fake a sore throat to get out of school. Chooks are much more like the Black Knight in *Monty Python and the Holy Grail*. 'It's only a flesh wound.' They'll fake being fine so adeptly that by the time you notice they're sick, it's often too late.

The trick is to pay close attention when you tend to them in the morning or evening. Is one of your gals keeping herself apart from the others? Is she sneezing? Are her eyes closed? Any sign that she's not acting entirely like a chicken (being very busy with important chicken business), you should separate her immediately from the other gals. Respiratory issues are highly contagious, and while diseases are less prevalent in backyard flocks with only a handful of chooks, it's best practice to whisk her away and isolate. There are some truly awful chicken viruses.

SKIRT: The amount of feathery underlay that a chicken can bring to hatch eggs. The larger the skirt, the more eggs she can sit on.

SPENT HEN: This is a charming (not charming) phrase used to describe a chook who has passed her laying peak. (You'll notice a lot of chook jargon relates to where the girls are at in regards to egg-laying.) The industry considers layer hens to be 'spent' at eighteen months old, which is when they're shipped off to the abattoir, and the egg farm, be it free-range or caged, is restocked with new point-of-lay hens. 'Spent' doesn't necessarily mean 'stopped laying entirely'.

SUNBATHING: Your chicken is not dead. It's sunbathing. The first time you see it, *yeek*, it looks like it's dislocated in seventeen places. It's disconcerting, but chooks *love* getting a bit of sun. It's good for them. Panic ye not.

TREATS: Use your common sense. Chickens are like people, if you let them they'll eat nothing but the fun things. Sweet corn, for example, is like chocolate for chickens. It can make them fat, and if you give it to them every day they will take it for granted *and* they'll fill up on it. You'll notice that if you give them grain, they'll pick out the sunflower seeds first. Their faves. But to get the balanced diet they need all of the grain. Their birthday doesn't have to be every day. There's a perception that chickens, left to their own devices, will miraculously choose what's best for them and, to a point, that's true. But if you give them a lot of corn, mealworms and tuna they'll begin turning their beaks up at healthier options. Just like people.

UTILITY BREEDS: Chickens bred to be utilised for their egg output, or their meat production. Farm chooks.

VETS: It's worth finding one who has experience with poultry. They're on the rise!

***VULPES VULPES*:** Latin for effing foxes.

WATER: Yes, they need it, and keep it fresh and cool. But I didn't come here to tell you that self-evident piece of genius. Here's something I learned the hard way: when you have newly hatched chicks, *do not* put water down for them. The chicks are a) idiots who will drown, and b) exceedingly well designed little units that contain enough of the egg yolk to supply their food and hydration needs for the first twenty-four hours. After they've been around a day, make sure water has pebbles in it up to the surface so they absolutely, 100 per cent cannot drown in it. Because, I can't stress this enough, they will drown in it. ●

ACKNOWLEDGEMENTS

A book – particularly a chicken book – doesn't just happen, and we would like to gratefully acknowledge the many people who have contributed to *This Chicken Life*. Firstly, our publisher, Mary Small at Plum, who saw the potential in the project, and the mighty team she assembled: project editor Jane Winning, copy editor Emma Schwarcz, and designer Michelle Mackintosh. Also to our agent, fellow chook enthusiast Jenny Darling, who was a champion.

We both owe our families much in the way of gratitude and thanks. Fiona to her partner Greg Craske, who fed and watered her as she sat writing at the computer for what seemed an eternity, and Ilana's husband Stuart and son Ari, who kept the home fires burning while Fiona and Ilana traversed Australia meeting beautiful chicken people, and didn't complain too much about their home being smothered in chook photographs. Ilana also thanks her beloved mum Shoshanna, and Chele, for the beginning of her chicken journey.

We are absolutely indebted to, and blown away by, all of the gorgeous, generous chicken-loving people we met around Australia. The adoration for chooks is real, and we were astounded by everyone's generosity of spirit, honesty and passion. It's been a privilege. Particular thanks to Megg and Jessamy Miller, whose chicken knowledge is bottomless, and who kindly acted as sounding boards and fact-checkers. Thanks also to Jessamy for generously sharing her house (and chooks!) for our cover shoot.

Other thanks go out to Sue Maslin, who introduced us to Jenny Darling; one must never underestimate the power of putting people together. There are also the many friends and colleagues who transcribed the interviews. Transcribing is a thankless job, so it's important we thank: Annie Rose Cosgriff, Danny Blay, Sophie Smyth, Claire Sullivan, Tim Hurley, Mark Liddell, Ryan Smedley, Nikki Spunde, Teresa Blake, Stephanie Adams, Claire Taylor, Maryanne Lynch, Kathryn Carter and Sandy Morrison.

Fiona would also like to thank her colleagues at RMIT, who put up with her distracted behaviour in the first half of 2019. And is indebted to the fierce 'you can do it!' cheerleading from her writer's group friends, Sian Prior and Mel Cranenburgh, and bestie Sonya Pemberton. Fiona would also like to thank her chickens, without whom none of this would have happened. They care not for thanks, of course, but would appreciate a few mealworms.

A PLUM BOOK

First published in 2019 by
Pan Macmillan Australia Pty Limited
Level 25, 1 Market Street,
Sydney, NSW 2000, Australia

Level 3, 112 Wellington Parade,
East Melbourne, Victoria 3002, Australia

Design by Michelle Mackintosh
Typesetting by Megan Ellis
Editing by Emma Schwarcz
Photography by Ilana Rose
Colour reproduction by Splitting Image
Colour Studio
Printed and bound in China by 1010 Printing
International Limited

A CIP catalogue record for this book is available
from the National Library of Australia.

10 9 8 7 6 5 4 3 2 1